The Ecphoras

In the Miocene and Pliocene fossil shell beds of the eastern United States, the single most spectacular molluscan species radiation is seen in the ecphora shells (the Tribe Ecphorini). These bizarrely shaped gastropods, with their distinctive ribbed shell sculpture, represent a separate branch of the Subfamily Ocenebridae, Family Muricidae. Characteristically, these muricid gastropods are heavily ornamented with spiral ribs and cords and are considered some of the most beautiful and interesting groups of fossil mollusks found along the Atlantic Coastal Plain and Floridian Peninsula. The ecphoras are greatly sought after by fossil collectors.

The ecphora faunas, and their individual species and subspecies, are illustrated and described in detail in *The Ecphoras: Iconic Fossils of Eastern North America,* along with photographs of ecphora-bearing geological units and in situ specimens. The authors list the 67 known species and subspecies that are recognized as valid, arranged by the eight genera and five subgenera that encompass these taxa

The Ecphoras

Iconic Fossils of Eastern North America

Edward J. Petuch
David P. Berschauer

CRC Press
Taylor & Francis Group
Boca Raton London New York

CRC Press is an imprint of the
Taylor & Francis Group, an **informa** business

First edition published 2023
by CRC Press
6000 Broken Sound Parkway NW, Suite 300, Boca Raton, FL 33487-2742

and by CRC Press
4 Park Square, Milton Park, Abingdon, Oxon, OX14 4RN

CRC Press is an imprint of Taylor & Francis Group, LLC

ISBN: 978-1-032-45601-0 (hbk)
ISBN: 978-1-032-45674-4 (pbk)
ISBN: 978-1-003-37818-1 (ebk)

DOI: 10.1201/9781003378181

Typeset in Times New Roman font
by KnowledgeWorks Global Ltd.

Dedication

This book is dedicated to the following:

Linda J. Petuch, Eric and Rasa Petuch,

Brian Petuch and Kendra Berentsen, and Jennifer Petuch

and

Felicia Weisbrot Berschauer, Morgan, Jeremy, Lincoln and Brantley Coker,

Jonathon, Tawni, Nora and Emmett Berschauer

THE ECPHORAS: ICONIC FOSSILS OF EASTERN NORTH AMERICA

TABLE OF CONTENTS

Ecphora (Powhatan) gardnerae unnamed subspecies, length 92 mm, from the base of the Eastover Formation near St. Inigoes Creek, St. Mary's County, Maryland. Specimen collected by the senior author and now in the collection of the Academy of Natural Sciences of Drexel University, Catalog No. ANSP IP 52868.

ACKNOWLEDGMENTS

Without the help and interest of the following individuals, this book on ecphoras could never have been produced. To all of them, we are greatly indebted for their assistance in field collecting, for allowing us access to their properties, and for the donation of valuable ecphora specimens. We thank the following: (for assistance with collecting in Maryland or for the donation of Maryland ecphora specimens) **Donald F. Asher; Norman Riker; Connie Smith; Amy Plummer; Joseph Turner; Robert Schmidt; Robert Shaw and the Chancellor Point Natural History Area; Irene White; Edward Mattingly; Dr. Bretton Kent; Dr. Elizabeth Dudley; Dr. Geerat Vermeij; David Bohaska and the Calvert Marine Museum; Wolfgang Grulke; Dr. Sylvard Kool; Linda Petuch; the Chapman Family;** (for assistance with collecting in Virginia or for the donation of Virginia ecphora specimens) **James and Bonnie Vautrot; Beth Permelia; Dahl Daniel; Eric Kendrew; Eddie Matchett; Dennis Sargent; Eleanor Marr;** (for assistance with collecting in North Carolina or for the donation of North Carolina ecphora specimens) **Ralph Chamness; I.K. ("Tex") Gilmore; Frank and Becky Hynes; Dr. Edward Gerrard; Rachel ("Rae") Germon;** (for the donations of specimens from the Duplin Formation at Eagle Point, South Carolina) **Michael Bruggeman; Gary Leonard;** (for assistance with collecting in Florida or for the donation of Florida ecphora specimens) **John Spengler; Richard Duerr; Charles and Violet Hertweck; Mary Mansfield; Mitchell Hope; Dale and Patricia Stream; Hugh Cannon and Quality Aggregates Inc.; Adam Nosel and the Ashland Petroleum and Asphalt Corporation (APAC); Brian Schnirel; Druid Wilson; Evelyn Bradley; Steve and Roxane Wilson.** For technical assistance, manuscript review, and editorial guidance when assembling this book, we thank **Alice Oven, Michele Dimont,** and **Shikha Garg,** CRC Press, Taylor and Francis Group.

Ecphora roxaneae dalli Petuch and Berschauer n. subsp., length 111 mm, Goose Creek Formation, South Carolina.

INTRODUCTION. The Ecphoras: Eastern North America's Iconic Fossils

In the Miocene and Pliocene fossil shell beds of the eastern United States, from southern New Jersey south to southern Florida, the single most spectacular molluscan species radiation is seen in the ecphora shells (vernacular usage for any member of the Tribe Ecphorini). These bizarrely-shaped gastropods, with their distinctive ribbed shell sculpture, represent a separate branch of the Subfamily Ocenebridae, Family Muricidae. Characteristically, these muricid gastropods are heavily ornamented with spiral ribs and cords and are considered to be some of the most beautiful and interesting groups of fossil mollusks found along the Atlantic Coastal Plain and Floridian Peninsula. Because of their exotic shell sculpture, large size, and intrinsic beauty, the ecphoras are greatly sought after by fossil collectors and are second only to shark teeth as prized collector's items in the Chesapeake Miocene formations of Maryland, Virginia, and northern North Carolina.

The intrinsic beauty and unusual appearance of ecphora shells have been known since the first European naturalists began exploring North America, and a specimen of an unidentified ecphora (probably *Ecphora quadricostata*) was the first American fossil gastropod to be illustrated in a natural history text (the 1770 Huddesford Edition of Martin Lister's *Historiae sive Synopsis Methodique Conchyliorum*). Since then, an additional 66 species have been discovered and named by several molluscan paleontologists, demonstrating that the ecphora shells were a more important and diverse part of the American paleofauna than was previously thought. Of the Atlantic Coastal Plain areas where there are exposures of sediments containing ecphoras, the Calvert Cliffs area of western Chesapeake Bay contains the largest number of collectible species and encompasses faunas from three separate geological formations: the **Calvert Formation**, the **Choptank Formation**, and the **St. Mary's Formation**. These formations and their component ecphoras are discussed, in detail, in the following chapters.

Although originally thought to be classic fossils for only the Miocene of the eastern United States, the ecphoras actually survived into the subsequent Pliocene Epoch and are well represented by a large number of species in Virginia, the Carolinas, and Florida. Indeed, the very last-living ecphora, *Latecphora hertwecki*, inhabited the late Pliocene **Myakka Lagoon System** off southwestern Florida, long after all the other ecphoras had become extinct elsewhere in eastern North America (see Petuch and Berschauer, 2021: 68-69, 121-131). Interestingly, the oldest-known ecphora, the Aquitanian Miocene *Ecphorosycon tampaensis*, was first discovered in southwestern Florida, the same area where the last ecphora species lived its final days. Southern Florida, then, can be seen to have been both the birthplace of the ecphoras and also the place of their death.

We here list the 67 known species and subspecies of ecphoras that are recognized as valid and which are discussed in the following chapters of this book. These are arranged by the eight genera and five subgenera that encompass these taxa.

Family Muricidae Rafinesque, 1815
Subfamily Ocenebrinae Cossmann, 1913
Tribe Ecphorini Petuch and Berschauer, new tribe
Genus *Ecphorosycon* Petuch, 1988
Ecphorosycon tampaensis (Dall, 1890)
Ecphorosycon pamlico (Wilson, 1987)
Ecphorosycon lindajoyceae (Petuch, 1993)
Ecphorosycon kalyx (Petuch, 1988)
Genus *Siphoecphora* Petuch, 1988
Siphoecphora aurora (Wilson, 1987)
Siphoecphora gerrardi new species
Genus *Chesathais* Petuch, 1988
Chesathais ecclesiasticus (Dall, 1915)
Chesathais whitfieldi Petuch, 1988
Chesathais pungoensis new species
Chesathais donaldasheri Petuch, 1988
Chesathais lindae Petuch, 1988
Chesathais drumcliffensis Petuch, 1988
Genus *Trisecphora* Petuch, 1988 sensu stricto
Trisecphora precursor new species
Trisecphora tricostata (Martin, 1904)
Trisecphora governorensis new species
Trisecphora scientistensis (Petuch, 1992)
Subgenus *Recurvecphora* Petuch and Berschauer, new subgenus
Trisecphora (Recurvecphora) recurvicostata new species
Trisecphora (Recurvecphora) schmidti (Petuch, 1988)
Trisecphora (Recurvecphora) martini (Petuch, 1988)
Subgenus *Eccentrella* Petuch and Berschauer, new subgenus
Trisecphora (Eccentrella) chamnessi (Petuch, 1988)
Trisecphora (Eccentrella) eccentrica (Petuch, 1988)
Subgenus *Matoaka* Petuch and Berschauer, new subgenus
Trisecphora (Matoaka) prunicola (Petuch, 1988)
Trisecphora (Matoaka) carolinensis (Petuch, 1988)
Trisecphora (Matoaka) bartoni Petuch and Drolshagen, 2010
Trisecphora (Matoaka) smithae (Petuch, 1988)
Trisecphora (Matoaka) patuxentia (Petuch, 1988)
Trisecphora (Matoaka) shattucki (Petuch, 1988)
Genus *Ecphora* Conrad, 1843 sensu stricto
Ecphora wardi Petuch, 1988
Ecphora mattinglyi Petuch, 2004
Ecphora sandgatesensis Petuch, 1988
Ecphora meganae Ward and Gilinsky, 1988
Ecphora amyae Petuch and Drolshagen, 2010

Ecphora kochi Ward and Gilinski, 1988
Ecphora roxaneae Petuch, 1991
Ecphora roxaneae dalli Petuch and Berschauer, new subspecies
Ecphora leecreekensis Petuch, 1988
Ecphora quadricostata (Say, 1824)
Ecphora quadricostata striatula Petuch, 1986
Ecphora quadricostata umbilicata (Wagner, 1839)
Ecphora quadricostata rachelae Petuch, 1988
Subgenus *Rhombecphora* Petuch and Berschauer, new subgenus
Ecphora (Rhombecphora) harasewychi Petuch, 1988
Ecphora (Rhombecphora) rikeri Petuch, 1988
Subgenus *Powhatan* Petuch and Berschauer, new subgenus
Ecphora (Powhatan) calvertensis Petuch, 1988
Ecphora (Powhatan) chesapeakensis Petuch, 1988
Ecphora (Powhatan) williamsi Ward and Gilinsky, 1988
Ecphora (Powhatan) conoyensis Petuch, 2004
Ecphora (Powhatan) asheri Petuch, 1988
Ecphora (Powhatan) germonae Ward and Gilinski, 1988
Ecphora (Powhatan) gardnerae Wilson, 1987
Ecphora (Powhatan) gardnerae angusticostata Petuch, 1988
Ecphora (Powhatan) grulkei new species
Ecphora (Powhatan) whiteoakensis Ward and Gilinski, 1988
Ecphora (Powhatan) whiteoakensis hopei Petuch, 1991
Ecphora (Powhatan) pachycostata Petuch, 1988

Genus *Planecphora* Petuch, 2004
Planecphora turneri (Petuch, 1992)
Planecphora vokesi (Petuch, 1988)
Planecphora choptankensis (Petuch, 1988)
Planecphora delicata (Petuch, 1988)
Planecphora hertweckorum (Petuch, 1987)
Planecphora mansfieldi (Petuch, 1988)

Genus *Globecphora* Petuch, 1994
Globecphora floridana (Petuch, 1988)
Globecphora floridana streami (Petuch, 1994)
Globecphora parvicostata (Pilsbry, 1911)
Globecphora mardieae new species

Genus *Latecphora* Petuch, 1988
Latecphora violetae (Petuch, 1988)
Latecphora bradleyae (Petuch, 1987)
Latecphora hertwecki (Petuch, 1988)

These beautiful and unusual murex shells are shown in the iconographies at the end of each of the following chapters. It is our hope that this book will serve as a field guide for amateur and professional paleontologists alike, and will give new insights into the evolutionary history and stratigraphy of eastern North America's most iconic group of fossil shells, the ecphoras. In the predecessor to this book, the 1988 *Field Guide to the Ecphoras*, a point of confusion as to the actual date of publication was inadvertently introduced. Projecting when the book might be published, the senior author stated that the date of publication was "15 February 1989". Since the book was actually published a few months earlier, in late 1988 (as stated in the front matter, "Copyright 1988 by the Coastal Education and Research Foundation"), all the taxa proposed in that monographic work should have publication dates of 1988, not 1989. This discrepancy in publication dates has been corrected in this subsequent volume.

Edward J. Petuch, Ph.D. and David P. Berschauer, J.D., 2022

View of a prime *Chesathais* collecting locality, along the Calvert Cliffs south of Governor Run, near St. Leonards, Calvert County, Maryland. Here, the highly-fossiliferous Shattuck Zone 18, Drumcliff Member of the Choptank Formation can be seen to overlay the darker and less-fossiliferous Shattuck Zone 17, St. Leonards Member of the Choptank Formation (at the base of the cliff). Eleven different ecphoras have been collected at this locality, including the most desirable species, the rare and beautiful rapaniform ecphora, *Chesathais lindae*.

CHAPTER 1. Systematics and Classification of the Ecphoras

The senior author (Petuch, 1988) originally considered the ecphoras to belong in their own extinct subfamily of the Muricidae (Murex Shells), the Ecphorinae. Because of the similarities in size and shell morphology, Bouchet *et al.* (2017) synonymized the Subfamily Ecphorinae with the Subfamily Ocenebrinae (the Ecphorini), as representing an extinct eastern North American clade of the worldwide subfamily. The living western North American genus *Forreria* appears to represent the closest group of extant ocenebrines to the extinct clade that we describe in the following section. This is especially apparent when considering that both *Forreria* and the ecphora shells have a two-layered mineralogical shell structure, with an inner layer of aragonite and an outer layer of calcite. The higher-order systematic classification scheme for our new clade, the Tribe Ecphorini, is arranged as follows:

Class Gastropoda Cuvier, 1797
Subclass Orthogastropoda Ponder and Lindberg, 1995
Superorder Caenogastropoda Cox, 1960
Order Sorbeoconcha Ponder and Lindberg, 1997
Infraorder Neogastropoda Wenz, 1938
Superfamily Muricoidea Rafinesque, 1815
Family Muricidae Rafinesque, 1815
Subfamily Ocenebrinae Cossmann, 1903
Tribe Ecphorini Petuch and Berschauer, new tribe

Description: (originally described as a subfamily of the Family Thaididae; now considered to be a tribe within the Subfamily Ocenebrinae, Family Muricidae; see Petuch, 1988: 29-30) "Shells inflated, globose, generally thickened and heavy; all genera with flaring, wide apertures and open umbilici; parietal shields either adherent **or** only partially attached to earlier whorl; siphonal canals generally elongated, well developed in most genera; almost all genera heavily sculptured with prominent raised cords or ribs, numbering three, four, five, or more; ribs may be simple and bladelike in form or complex, with flaring edges, producing 'T'-shaped cross section."

Morphological Characters of the Ecphorini

Shells of the Ecphorini characteristically are composed of two distinct calcium carbonate layers, with an outer layer of resistant calcite and an inner layer of softer aragonite. The outer calcite layer is frequently stained red, brown, or yellow and is often extremely well preserved. The inner aragonite layer, exposed within the aperture, is white in color but is often dissolved away in specimens from leached sedimentary layers. The 67 known species and subspecies of ecphoras cover a wide variety of morphological types and these taxa can be distinguished from each other by the examination and comparison of the following eight shell characters.

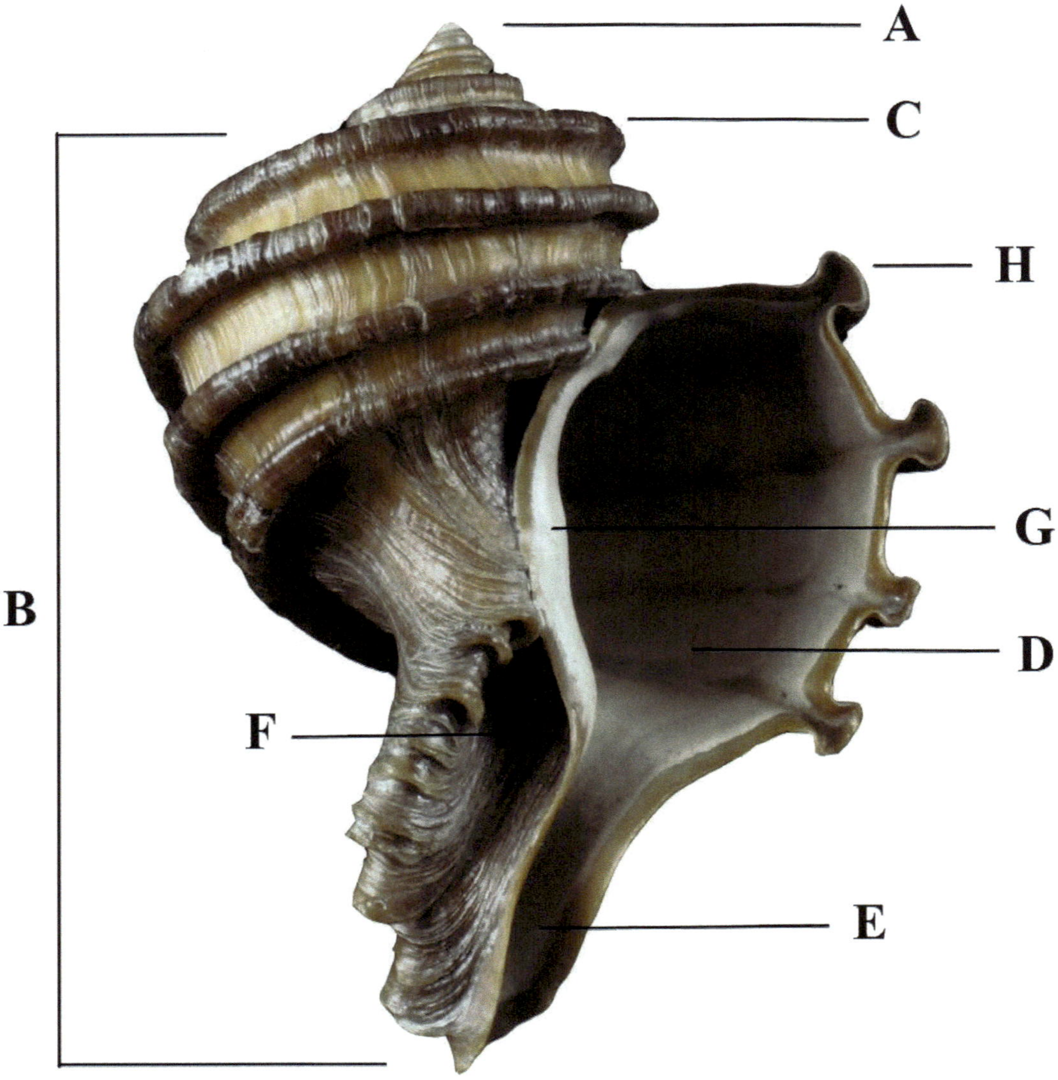

ILLUSTRATION OF A TYPICAL ECPHORA SHELL AND ITS PARTS
A= apex; **B**= body whorl; **C**= shoulder; **D**= aperture; **E**= siphonal canal; **F**= umbilicus; **G**= parietal wall; **H**= rib.

These important morphological characters are:

1. SHELL SHAPE: The shape of the shells, particularly that of the shell's body whorl, differs greatly between the various genera and subgenera of ecphoras. Some, such as the five-ribbed *Globecphora floridana*, are highly inflated and globose, often being the shape and size of a cantaloupe or small melon. Other species, such as *Ecphora mattinglyi*, are highly elongated and protracted, with extremely protracted and fragile spire whorls. The main types of ecphora shell shapes include the following, with an example of each form:

- **Globose** (*Globecphora floridana*); shells highly inflated with greatly expanded body whorls.

- **Cylindrical** (*Ecphora sandgatesensis*); shells round in cross section but straight-sided and narrow.
- **Pyriform (Pear-Shaped)** (*Siphoecphora aurora*); shells with rounded body whorls and elongated siphonal canals.
- **Vermiform** (*Trisecphora (Eccentrella) eccentrica*); shells extremely protracted and wormlike, with uncoiled, detached spire and body whorls.
- **Fusiform** (*Chesathais lindae*); spindle-shaped shells with high spires, narrow body whorls, and elongated siphonal canals.
- **Vase-Shaped (Vasiform)** (*Trisecphora (Matoaka) bartoni*); shells with broad shoulder areas, tapering quickly to the anterior end, producing a vase-like shape.
- **Turbinate** (*Latecphora bradleyae*); shells broad and inflated, with rounded shoulders and low spire whorls, having the shape of an inverted cone.

2. SHELL SIZE: The range in size among the ecphoras is substantial, with some species, such as *Ecphorosycon tampaensis*, being very small and delicate and others, such as the giant *Ecphora (Powhatan) gardnerae*, reaching the size of a small melon. The range of sizes, over the entire evolutionary history of the Tribe, extends from 15-20 mm for *E. tampaensis* to over 150 mm for *E. (Powhatan) gardnerae*. The older, early Miocene species are generally quite small, averaging lengths of only around 30 mm, with the only exception being the large pyriform genera, *Siphoecphora* and *Ecphorosycon*, which can reach lengths of 75 mm. The late Miocene and Pliocene ecphoras all are much larger than their early Miocene ancestors, with most taxa averaging over 90 mm in length.

3. SPIRE HEIGHT: The spire whorls differ greatly between genera and subgenera, with some species having low, nearly flattened spire whorls, such as *Globecphora parvicostata* and *Ecphora (Powhatan) pachycostata*, while others have greatly elevated and protracted spire whorls, such as *Ecphorosycon kalyx* and *Chesathais whitfieldi*. These differences in spire height and structure are important diagnostic features for separating subgenera and species complexes.

4. RIB COUNT: Although the general perception of ecphoras, by collectors and non-malacologists, is that they are four-ribbed shells, all resembling the well-known *Ecphora quadricostata*, a detailed review of the family shows that rib count is specific to each genus and subgenus. The six types of rib counts include the following (with an example of a species for each count):

- **Multiple Cords** (*Chesathais ecclesiaticus*); shells with as many as eight main cords on the body whorl, of variable size, with some being larger than others.
- **Three Main Ribs** (*Trisecphora (Matoaka) carolinensis*); shells with only three equal-sized ribs around the body whorl and no other primary ornamentation.
- **Three Main Ribs with Fourth Smaller Rib** (*Trisecphora tricostata*); shells with three main ribs around the body whorl and a fourth smaller subsidiary cordlike rib around the siphonal canal-body whorl juncture.
- **Four Main Ribs** (*Planecphora choptankensis*); shells with only four equal-sized ribs around the body whorl.

- **Four Main Ribs with Fifth Smaller Rib** (*Ecphora (Powhatan) whiteoakensis hopei*); shells with four main ribs around the body whorl and a fifth smaller subsidiary cordlike rib around the siphonal canal-body whorl juncture.
- **Five Main Ribs** (*Ecphora (Powhatan) asheri*); shell with five equal-sized ribs.

5. RIB STRUCTURE: The shape of the ribs is the single most important morphological criterion for differentiating the species, genera, and subgenera of ecphoras. This character is relatively invariant, with each taxon consistently having a set type of rib and showing little variation. The rib types include:

- **Rounded Cord** (*Globecphora floridana*); ribs that are directly adherent to the body and spire whorls, only slightly raised and rounded in cross section.
- **Low Cord with Flattened Edge** (*Ecphora (Rhombecphora) rikeri*); ribs that are directly adherent to the body and spire whorls, slightly raised, with a flattened edge and rectangular in cross section.
- **Thin Blade with Rounded Edge** (*Planecphora mansfieldi*); ribs that are very thin, delicate, and bladelike, often noticeably projecting outward from the body whorl and with rounded edges.
- **Thin Rib with Upturned, Cuplike Edge** (*Trisecphora (Recurvecphora) schmidti*); ribs that are thin and bladelike, but curl upward (posteriorward), producing a distinct trough-like channel.
- **Thin Rib, Slightly "T"-Shaped** (*Ecphora quadricostata*); ribs that are thickened, with distinctly flattened edges that are slightly "T"-shaped in cross section.
- **Thick Rib, Distinctly "T"-Shaped** (*Ecphora (Powhatan) gardnerae*); ribs that are thickened, with wide flanges on the edges and are strongly "T"-shaped in cross section. Ribs of this type frequently are sculpted with one or two deeply-incised cords.
- **Wide Rib, Extremely "T"-Shaped** (*Latecphora bradleyae*); ribs that are extremely wide and flattened, with very wide flanges on the rib edges; on advanced forms, the edges are almost in contact and the rib is broadly "T"-shaped. Ribs of this type frequently are sculpted with two, three, or four deeply-incised cords.

6. SHELL SCULPTURE: This morphological character refers to the texture of the shell surface that is exposed between the ribs and cords. These intracostal areas can be smooth and polished (as on *Globecphora parvicostata*); sculpted with fine spiral threads, which give the shell a silky appearance (as on *Ecphora wardi*); ornamented with numerous fine spiral cords and threads (as on *Ecphora quadricostata striatula*); or ornamented with very strong spiral ribs, some of which may be almost as large as the main ribs (as seen on *Chesathais drumcliffensis*). The shell sculpture is very consistent within any given species or subspecies and can be used to determine individual taxa.

7. SIPHONAL CANAL STRUCTURE: Both the length and the structure of the siphonal canal in the Tribe Ecphorini is variable between genera and subgenera. Some genera, such as *Siphoecphora* and *Ecphorosycon*, have very long, narrow siphonal canals that approximate those seen in the gastropod Family Busyconidae and the living Eastern

Pacific ocenebrine muricid genus *Forreria*. Other genera, such as *Latecphora* and *Globecphora*, have proportionally very short and broad canals, resembling those seen on smaller ocenebrine groups such as *Ocenebra*, *Ocenebrina*, and *Urosalpinx*. Genera such as *Ecphora* and *Planecphora* have siphonal canals that fall between these two extremes; being intermediate in both length and width.

8. STRUCTURE AND FORM OF THE UMBILICUS: The umbilical area, adjacent to the aperture, also varies in shape, size, and structure. Some genera, such as *Latecphora* and *Globecphora*, have extremely wide, open, and flaring umbilici while other genera, such as *Siphoecphora* and *Chesathais*, have very narrow, slot-like umbilici. The genera *Ecphora* and *Trisecphora* fall between these two extremes in umbilical form.

These eight morphological criteria, when studied and applied in consort, demonstrate that the Tribe Ecphorini is actually much more species-rich than previously thought and that its 67 species and subspecies belong to eight separate higher taxa. These include the following genera:

Ecphorosycon (Figure 1.1 A)
Siphoecphora (Figure 1.1B)
Chesathais (Figure 1.1C)
Trisecphora (Figure 1.1D)
Ecphora (Figure 1.2A)
Planecphora (Figure 1.2B)
Globecphora (Figure 1.2C)
Latecphora (Figure 1.2D)

Five evolutionary offshoots (subclades) are also known and these are described here as subgenera of *Trisecphora* and *Ecphora*. These include:

Eccentrella (subgenus of *Trisecphora*; Figure 1.3A)
Recurvecphora (subgenus of *Trisecphora*; Figure 1.3B)
Matoaka (subgenus of *Trisecphora*; Figure 1.3C)
Rhombecphora (subgenus of *Ecphora*; Figure 1.3D)
Powhatan (subgenus of *Ecphora*; Figure 1.3E)

All of these genera and subgenera are described in the following chapters of this book.

Figure 1.1 Genera of the Tribe Ecphorini.
A= ***Ecphorosycon* Petuch, 1988** (*Ecphorosycon pamlico* (Wilson, 1987), length 74 mm); **B= *Siphoecphora* Petuch, 1988** (*Siphoecphora aurora* (Wilson, 1987), length 70 mm); **C= *Chesathais* Petuch, 1988** (*Chesathais ecclesiasticus* (Dall, 1915), length 50 mm); **D= *Trisephora* Petuch, 1988** (*Trisecphora tricostata* (Martin, 1904), length 35 mm).

Figure 1.2 Genera of the Tribe Ecphorini.
A= ***Ecphora* Conrad, 1843** (*Ecphora quadricostata* (Say, 1824), length 60 mm); **B= *Planecphora* Petuch, 2004** (*Planecphora mansfieldi* (Petuch, 1988), length 85 mm); **C= *Globecphora* Petuch, 1994** (*Globecphora floridana* (Petuch, 1988), length 110 mm); **D= *Latecphora* Petuch, 1988** (*Latecphora bradleyae* (Petuch, 1987), length 93 mm).

Figure 1.3 Subgenera of the Tribe Ecphorini.
A= ***Eccentrella*** **new subgenus of** ***Trisecphora*** (*Trisecphora (Eccentrella) eccentrica* Petuch, 1988), length 56 mm); **B=** ***Recurvecphora*** **new subgenus of** ***Trisecphora*** (*Trisecphora (Recurvecphora) schmidti* (Petuch, 1988), length 30 mm); **C=** ***Matoaka*** **new subgenus of** ***Trisecphora*** (*Trisecphora (Matoaka) patuxentia* (Petuch, 1988), length 37 mm); **D=** ***Rhombecphora*** **new subgenus of** ***Ecphora*** (*Ecphora (Rhombecphora) rikeri* Petuch, 1988, length 80 mm); **E=** ***Powhatan*** **new subgenus of** ***Ecphora*** (*Ecphora (Powhatan) gardnerae* Wilson, 1987, length 120 mm).

Ecology of the Ecphora Shells

Since the entire Tribe Ecphorini is extinct, the ecologies of the various genera can only be partially reconstructed from inferences in the fossil record. As in all the living members of the muricid Subfamily Ocenebrinae, ecphoras were carnivorous gastropods that fed by drilling holes in the shells of other mollusks, mostly bivalves, and rasping out the exposed flesh. This feeding style is readily seen in large bivalves collected from the St. Mary's Formation (Tortonian Miocene) along the St. Mary's River in Maryland. Many of the single valves of the large venerid clams *Mercenaria tetrica* and *Mercenaria mortoni* are found with over-sized holes drilled near the hinges. Some of these holes are over 5 mm in diameter, and are all of the typical muricid-type, straight-sided form. With the exception of one large muricid, *Ecphora (Powhatan) gardnerae*, all the other St. Mary's muricoideans, which include species of *Mariasalpinx*, *Urosalpinx*, *Scalarispira*, *Chesatrophon*, and *Laevityphis* are small, under 25 mm in length, and were incapable of drilling a 5 mm-diameter hole in a thick *Mercenaria* valve. This leaves only one possibility; *Ecphora (Pohatan) gardnerae* was the culprit and was the apex predator on large bivalves such as *Mercenaria*, *Dosinia*, and *Dallarca* (see Petuch and Drolshagen, 2010: 41-43: figure 3.3).

While the St. Mary's Formation contained only one species of ecphora that had virtually no competition for its food resource, the older Calvert and Choptank Formations (Langhian and Serravallian Miocene) contain molluscan assemblages with as many as five contemporaneous and sympatric ecphoras. A good example is Shattuck Zone 19, Boston Cliffs Member of the Choptank Formation, where five species occur together. To avoid competition, these species evolved different-sized shells and accordingly, preyed upon different size classes of bivalves. Small species, such as *Ecphora amyae* and *Trisecphora (Matoaka) delicata*, probably preyed upon small, thin-shelled bivalves such as *Aligenia*, *Diplodonta*, *Abra*, and *Paramya*. Larger ecphora species, such as *Ecphora (Powhatan) williamsi* and *Chesathais drumcliffensis*, would have had an abundance of large bivalve prey, including *Mercenaria capax*, *M. plena*, and *Marvacrassatella marylandica*, as these large clams form solid beds in Shattuck Zone 19; attesting to the immense biomass available to the resident ecphoras. The five Shattuck Zone 19 ecphoras make up the bulk of muricoidean predation, since only one small species of *Urosalpinx* and one *Laevityphis* are found along with the ecphoras.

While living on soft substrates on the open sea floor, and while feeding on shallow infaunal bivalves, the ecphoras were, themselves, vulnerable to molluscivore predators. Primary among these were crabs of the genera *Menippe* (Stone Crabs) and *Calappa* (Box Crabs), who used their claws to peel back the shell from the aperture and expose the unprotected snail. Ecphoras with large healed breaks are common (see Figure 1.4 A, B) and testify to the high frequency of crab attacks. Ecphoras were also often victims of drilling predation, mostly from Moon Snails of the family Naticidae. Conically-shaped, beveled naticid drill holes are commonly seen on ecphora shells, such as the one shown here on Figure 1.4 C. Since ecphoras lived on soft, sandy or muddy sea floors, which are the ideal habitats for large burrowing naticids, they were especially vulnerable to Moon Snail drilling. In the northern areas, naticids of the genera *Euspira* and *Glossaulax* where the principal drillers of ecphoras, while in southern areas, such as Florida, the genera *Naticarius* and *Glossaulax* were the primary drilling predators.

Figure 1.4 Examples of Predation on Ecphoras.
A= *Trisecphora (Matoaka) bartoni*, with massive shell damage due to a failed predation attack by a *Calappa* crab; the entire siphonal canal had been broken off and later regrown, leaving large healed scars and shell deformation; **B=** *Ecphora (Powhatan) pachycostata*, with large "C"-shaped fractures due to a failed attempt at peeling back the outer lip by a *Menippe* crab; the thickened ribs saved the snail and kept the crab from peeling the shell any farther back, but the attack left large healed scars from the extensive shell damage; **C=** *Globecphora parvicostata*, with a large drill hole in the middle of the body whorl, caused by predation from a large naticid Moon Snail, probably *Euspira* cf. *rogersi*; the hole is beveled, showing that it was produced by a moon snail and not by other muricid gastropods. Another failed attempt at drilling can be seen farther back on the body whorl, as evidenced by the shallow circular depression at approximately the same position as the successful drill hole that ended this ecphora's life.

Patterns of Evolution and Extinction in the Ecphoras

As discussed in Chapter 3, the oldest-known ecphora is the small *Ecphorosycon tampaensis*, which first appeared in the early Aquitanian Miocene and spread along the eastern North American coast from Florida and North Carolina, northward to Maryland. Various "ecphoras" that have been collected in older, Oligocene, formations along the east coast, such as the "*Ecphora* species" shown by Ward (1985: plate 6, figure 12) and the "*Ecphora wheeleri*" of Carter et al. (1994), have been subsequently shown to be small rapanine muricids, without a thick calcite outer layer. Some of these Oligocene-early Aquitanian Miocene ecphora-like rapanine muricids, such as "*Rapana*" *vaughani* Mansfield, 1937 of the Chattahoochee Formation, were placed in a new ecphorine genus, *Rapanecphora*, by the senior author (1988: 33-37). This polyphyletic genus, however, has subsequently been found to represent members of the Subfamily Rapaninae and not the subfamily Ocenebrinae, so these species will not be included in this book. The genus *Rapanecphora* should only be applied to the Oligocene *vaughani* from northern Florida. The senior author also included another small rapanine muricid, *Rapana biconica* Dall, 1915, under the Tribe Ecphorini, in this case in the genus *Chesathais* (Petuch, 1988: 112). Likewise, subsequent study showed that this distinctive Aquitanian Miocene shell, from the Tampa Member of the Arcadia Formation, is not an ecphora at all and, for that reason, is not included in this book. Although these Oligocene and earliest Miocene species bear a close resemblance to some ecphora groups like *Chesathais* and *Siphoecphora*, they do not appear to be related to the Tribe Ecphorini.

Over the entire lifespan of the Tribe Ecphorini, several pulses of evolution and extinction occurred, at times resulting in large species radiations that were later followed by impoverished faunas containing only one species. The first of these evolutionary pulses occurred during the Langhian Age of the early Miocene, when whole suites of new taxa appeared very rapidly in the fossil record. In Maryland, genera such as *Trisecphora*, *Chesathais*, and *Ecphora* coexisted with the older, archaic genus *Ecphorosycon* in the Calvert Formation. The peak time for this Langhian evolutionary explosion was during the deposition of the Plum Point Member (Shattuck Zone 10), when Maryland was at its warmest during the early Miocene. The subtropical marine conditions during Plum Point Member time are evidenced by the presence of tropical gastropod groups in Shattuck Zone 10, including *Xenophora* (*X. lindae* Petuch and Drolshagen, 2010), *Americoliva*, and *Hespererato*. The biodiversity during Plum Point Member time was very high (see Petuch and Drolshagen, 2010) and the warm water conditions supported a large fauna of small bivalve mollusks, offering a plethora of food resources for the predatory ecphoras. This resulted in the evolution of the largest ecphora fauna known from any single locality, with eight sympatric species being found at Plum Point alone.

The subsequent early Serravallian Age, unlike the warm climatic conditions of the late Langhian, was a time of cold conditions ("climatic minimum") on a global scale. Oceanic temperatures in Maryland, Virginia, and North Carolina dropped well below those necessary to support subtropical molluscan faunas and a mass extinction of tropical groups took place. Some of the classic tropical gastropod genera and families that disappeared abruptly from the Maryland fossil record at the beginning of the Serravallian Age, and never returned, include the following: *Amaea* (Epitoniidae); *Americoliva* (Olividae-Olivinae); *Xenophora* (Xenophoridae); *Hespererato* (Eratoidae); *Ficus* (Ficidae); *Sassia* (Personiidae); *Niso* (Melanellidae); *Architectonica* (Architectonicidae); and *Persicula* (Cysticidae) (see Martin, 1904; Petuch and Drolshagen, 2010; and Petuch, 2012, for illustrations and stratigraphic ranges of all of these taxa). These mid-Miocene global cooling events, and their associated mass extinctions, have been correlated by the senior author (Petuch, 2012) with catastrophic asteroid impacts, such as the early Serravallian Ries-Steinheim Impact in Germany (Shoemaker and Chao, 1961) and the mid-Serravallian Ewing and Puffin Impacts (Abbott *et al.*, 2003; Leung *et al.*, 2004; Gorter and Bayford, 2000). The dust clouds from these immense impacts cooled the Earth's climate dramatically, causing rapid glacial build-up and drops in sea levels. Surprisingly, the ecphoras appeared to be unaffected by the Serravallian cold times and flourished as the primary group of large muricids until the end of Serravallian time; producing entire complexes of new species in a second wave of ecphora evolution.

When the Earth entered a period of global warming ("climatic optimum") during the subsequent Tortonian Age, the polar ice caps and high-latitude glaciers melted, raising sea levels and flooding the coastal plain areas with warmer water conditions (see Oleinik et al., 2008). At the Serravallian-Tortonian boundary (Choptank-St. Mary's formational boundary, Shattuck Zones 19 and 20), a mass extinction of the ecphoras took place in the Chesapeake Miocene. This extinction is very obvious when comparing the ecphora fauna of the Boston Cliffs Member (Shattuck Zone 19), containing the genera *Trisecphora (Matoaka)*, *Ecphora*, *Ecphora (Powhatan)*, *Chesathais*, and *Planecphora*, with that of the subsequent Conoy Member (Shattuck Zone 20) with its single species (*Ecphora (Powhatan) conoyensis*). From the early Tortonian onward, only one species of *Ecphora (Powhatan)* is known to occur in any given geological formation in Maryland and Virginia. These are the classic large ecphoras that are found in the subsequent Shattuck Zones 21, 23, 24, 25 of the St. Mary's Formation and the two members of the Eastover Formation. This one lone subgenus of ecphora appears to have survived the Tortonian warming time, with its accompanying invasion of small muricid gastropods from the south, by being the largest drilling gastropod predator. This size advantage would have allowed *Powhatan* to have unimpeded access to large bivalve prey, such as the venerids *Mercenaria*, *Dosinia* and *Callista*; thus avoiding competition with its smaller muricid relatives such as *Urosalpinx*, *Mariasalpinx*, *Chesatrophon*, *Laevityphis*, *Lirosoma*, and *Scalaspira*, all of which are abundant in the St. Mary's Formation (see Petuch and Drolshagen, 2010 for illustrations of the St. Mary's muricid fauna).

The Tribe Ecphorini remained an impoverished group of muricids throughout the late Miocene, being represented by only *Ecphora (Powhatan)* in Maryland, Virginia, and North Carolina and *Globecphora* and *Ecphora (Powhatan)* in southern Florida (in the

Bayshore Formation; Petuch and Berschauer, 2021). At the end of the warm Messinian Age, the Earth was again struck by a large asteroid, this time in the Pamir Mountains of Tajikistan (the Kara-Kul Impact; Gurov and Gurova, 1993). The dust and aerosols launched into the atmosphere plunged the Earth into another cold climatic minimum, causing a mass extinction of marine organisms. The rapid lowering of oceanic temperatures and extinctions caused by this terminal Messinian impact, however, had no affect on the Tribe Ecphorini, which was already at its lowest biodiversity since the Aquitanian. The extinction of competitive muricid genera, such as *Zulloia* and *Calusathais* in Florida (Petuch and Berschauer, 2021), however, allowed the one or two surviving ecphoras to speciate rapidly and undergo a third wave of evolution in the subsequent early Zanclean Age of the Pliocene. During the cold times of the Serravallian and early Tortonian Ages, Florida appears to have acted as a refugium for the genera *Ecphora s.s.* and *Planecphora*, as these groups disappeared from the fossil record of Maryland and Virginia but made a re-appearance in the earliest Zanclean Murdock Formation of southern Florida. These relictual taxa had spread northward by mid-Zanclean time, combining with the impoverished Chesapeake faunas and producing a much richer assemblage that contained several genera, including *Ecphora s.s.*, *Ecphora (Powhatan)*, *Globecphora*, and *Planecphora*.

By the beginning of the Piacenzian Age of the Pliocene, marine climatic conditions had warmed to full tropical levels, as evidenced by the formation of immense coral reef complexes in Florida (the "Everglades Pseudoatoll"; see Petuch and Berschauer, 2021). At this time, there was an explosion of muricid genera and species within the tropical seas of the southeastern United States, including large genera such as *Phyllonotus*, *Hexaplex*, and *Chicoreus* (see Petuch, 1994 and Petuch and Berschauer, 2021 for illustrations of the hundreds of species of muricids found within the Everglades Pseudoatoll region). These fast-moving and active drilling murex shells would have out-competed the slower and more primitive ecphoras for their bivalve prey. The tropical water conditions would have also exposed the surviving ecphora taxa to increased predation by peeling and crushing crabs, as evidenced by the early Piacenzian extinction of the genera *Globecphora* and *Ecphora (Powhatan)*. In response to this increase in crab and crustacean predation, a new group of ecphoras made its appearance at this time, the wide-ribbed genus *Latecphora*. The extremely-expanded ribs of this genus would have created "I"-beam-shaped rib structures, in essence producing a double shell without doubling the shell weight. By the mid-Piacenzian Age, only two ecphora genera existed; the genus *Ecphora s.s.* and the newly-evolved *Latecphora*, with its anti-crab predation double shell. The last-living ecphora was a dwarf *Latecphora* found in the Tamiami Formation of southern Florida, which apparently succumbed to the combination of competition from other muricids and the heavy predation by tropical crabs.

Deposition of the Type Material of New Ecphora Taxa

Throughout this book, we describe eight new species and subspecies in context with their respective genera and subgenera. The holotypes and paratypes of these new taxa, and also those named previously by the senior author, are deposited at several different museums and bear different catalog numbers and institutional abbreviations. These are used throughout this book and include:

USNM= Department of Paleobiology, National Museum of Natural History, Smithsonian Institution, Washington, D.C.
AMNH FI= Fossil Invertebrate collection of the Division of Paleontology, American Museum of Natural History, New York City, New York
PRI= Paleontological Research Institute, Ithaca, New York
UF= Invertebrate Paleontology collection of the Florida Museum of Natural History, University of Florida, Gainesville, Florida
ANSP= Academy of Natural Science of Philadelphia, Philadelphia, Pennsylvania (now Drexel University)
MCZ= Museum of Comparative Zoology, Harvard University, Cambridge, Massachusetts
CM= Carnegie Museum of Natural History, Pittsburgh, Pennsylvania

Ecphora (*Powhatan*) cf. *whiteoakensis* - length 83 mm; a possible unnamed species or subspecies from the Eastover Formation at Aurora, North Carolina.

CHAPTER 2. Ecphoras as Stratigraphic Index Fossils

The geologic formations of the eastern United States, particularly the Chesapeake Bay area, contain a detailed record of the oceanographic history of the Miocene Epoch. The shell beds and geologic units of coastal Maryland, Virginia, and northern North Carolina were deposited within two large oceanic systems; the **Salisbury Sea** and the **Albemarle Sea** (Petuch, 2004: 7-11; Petuch and Drolshagen, 2010: 1-20). Although referred to previously as "embayments" (Gibson, 1983; Ward, 1985, 1992), these systems were far more complex than simple Atlantic coastal bays and represented true seas in the oceanographic sense. Although geographically small, the Salisbury and Albemarle Seas fit all the criteria for designations as true seas: they were bodies of salt water; they occupied geologically well-defined basins and were structurally bound on three sides; they each contained their own unique configurations of currents and water masses; and they contained their own characteristic endemic organisms and ecosystems. The coastal cliffs of Maryland and Virginia are, essentially, exposures of cross sections through the sea floors of the Salisbury and Albemarle Seas and contain the oceanographic history of the entire Miocene Epoch. Collecting areas for ecphora bearing units are shown here on pages 17 and 18.

As major global warming and cooling events took place during the Miocene (discussed in Chapter 1), the Salisbury and Albemarle basins alternately flooded and dried out as sea levels fluctuated. During warm transgressive intervals, thick, highly fossiliferous sea floors were deposited, and the remnants of these are the geological formations and members that are exposed in the cliffs of the Chesapeake Bay area. As the oceans retreated during cold regressive intervals, deposition ceased and the exposed sea floors were subject to erosion by rivers and streams, producing unconformities and gaps in the oceanographic history of Chesapeake Bay. The separate flooding and depositional episodes of each basin, bounded above and below by unconformities, are referred to as **subseas** (Petuch, 2004). During the Miocene, the basins of the Salisbury and Albemarle Seas, together, contained seven separate subseas. The early geologists and paleontologists, such as Martin (1904) and Shattuck (1904), recognized these five separate sets of sea floors and described them as distinct geologic formations and faunas.

The oceanography and geomorphology of each of these subseas where given by Petuch (2004) and Petuch and Drolshagen (2010) and the resident ecphora faunas were described. Based on these earlier works, we now recognize the following Miocene subsea divisions, including: (Salisbury Sea) **Old Church Subsea**, **Calvert Subsea**, **Patuxent Subsea**, and **St. Mary's Subsea**; and (Albemarle Sea) **Silverdale Subsea**, **Pamlico Subsea**, and **Rappahannock Subsea**. The Salisbury Sea was completely filled with sediments by the beginning of the Pliocene and ceased to exist as a recognizable oceanographic unit. At that time, the focus of marine deposition shifted southward to Virginia and North Carolina and the Albemarle Sea became the only major paleosea, extending inland as far west as Richmond, Virginia. The Pliocene subsea divisions include: (Albemarle Sea) **Williamsburg Subsea** and the **Yorktown Subsea**. Farther south, along southern North Carolina and northern South Carolina, another smaller Pliocene paleosea had developed by Zanclean time; the **Charleston Sea**. The ecphora-bearing formations of the Charleston Sea were deposited within the **Santee Subsea** and the **Duplin Subsea**. Two more southern Pliocene paleoseas also housed rich faunas of ecphoras; the **Choctaw Sea** of northwestern Florida and southwestern Georgia and the **Okeechobean Sea** of southern Florida. The Choctaw and Okeechobean Seas, each contained only one ecphora-bearing subsea during the early Pliocene, the **Jackson Subsea** (Choctaw Sea) and the **Tamiami Subsea** (Okeechobean Sea). The geography, oceanography, and molluscan faunas of each of all these Pliocene seas and subseas were described and illustrated in Petuch, 2004, and Petuch and Berschauer, 2021.

Stratigraphy of Ecphora-Bearing Units

Ecphoras are both distinctive and locally abundant and can be used by stratigraphic geologists as guide (index) fossils for the Miocene and Pliocene formations of the coastal southeastern United States, from Chesapeake Bay south to Florida. Since ecphoras underwent rapid evolution during the early Neogene, every species and subspecies is restricted to a single geologic unit, often to single numbered beds within depositionally-complex formations. A classic example of this type of geochronological sequestering is seen in the "Shattuck Zones" of the Maryland Miocene, where 24 numbered units in three geological formations are exposed in the cliffs along Chesapeake Bay and the Patuxent, St. Mary's, and Potomac Rivers (Shattuck, 1904). The eight morphological criteria outlined in Chapter 1 can be used to identify any Maryland ecphora species or subspecies and to differentiate it from related taxa that are subjacent (stratigraphically-lower) or suprajacent (stratigraphically-higher). Once the identity has been accurately determined, the layer from which the ecphora was found can be dated with certainty, using the stratigraphic positions and relative dating patterns that we show here in each of the following sections. This stratigraphic overview is divided into six sections, depending on epoch and geography. These include: the **Miocene of Maryland**; the **Miocene of Virginia and the Carolinas**; the **Pliocene of Virginia and the Carolinas**; the **Miocene of southern Florida**; the **Pliocene of southern Florida**; and the **Jackson Bluff Formation** of northwestern Florida.

Some of the more important localities where ecphoras have been collected on a regular basis are shown on the following two maps.

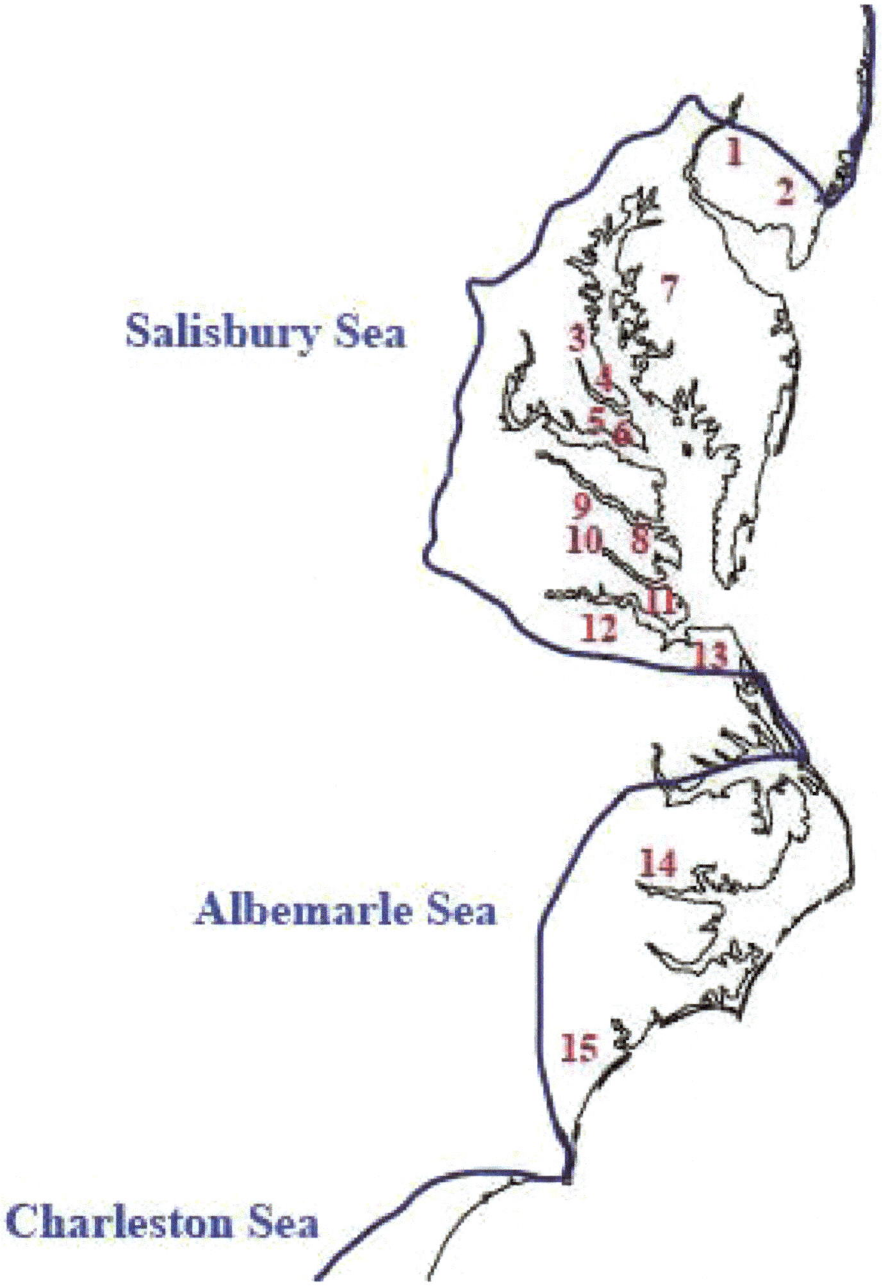

Map of the eastern United States showing some important northern localities where ecphoras have been regularly collected, including:

1= Jericho, New Jersey (in quarries)
2= Shiloh, New Jersey (in quarries)
3= Chesapeake Beach, Maryland (exposed in cliffs)
4= Plum Point, Scientist's Cliffs, Calvert Cliffs, and Cove Point, Maryland (exposed in cliffs)
5= Patuxent River, Maryland (exposed in cliffs)
6= St. Mary's River, Maryland (exposed in low cliffs)
7= Southeast Creek, at Church Hill, Maryland (exposed in low cliffs)
8= Rappahannock River, at Urbanna, Virginia (exposed in cliffs)
9= Mattaponi River, Virginia (exposed in low cliffs)
10= Pamunkey River, Virginia (exposed in low cliffs)
11= York River, at Yorktown, Virginia (exposed in cliffs)
12= James River, at Rushmere, Virginia (exposed in low cliffs)
13= James River, Norfolk area, Virginia (exposed in low cliffs)
14= Pamlico River, at Aurora, North Carolina (in quarries)
15= Magnolia, North Carolina (in stream cuts and sink holes)

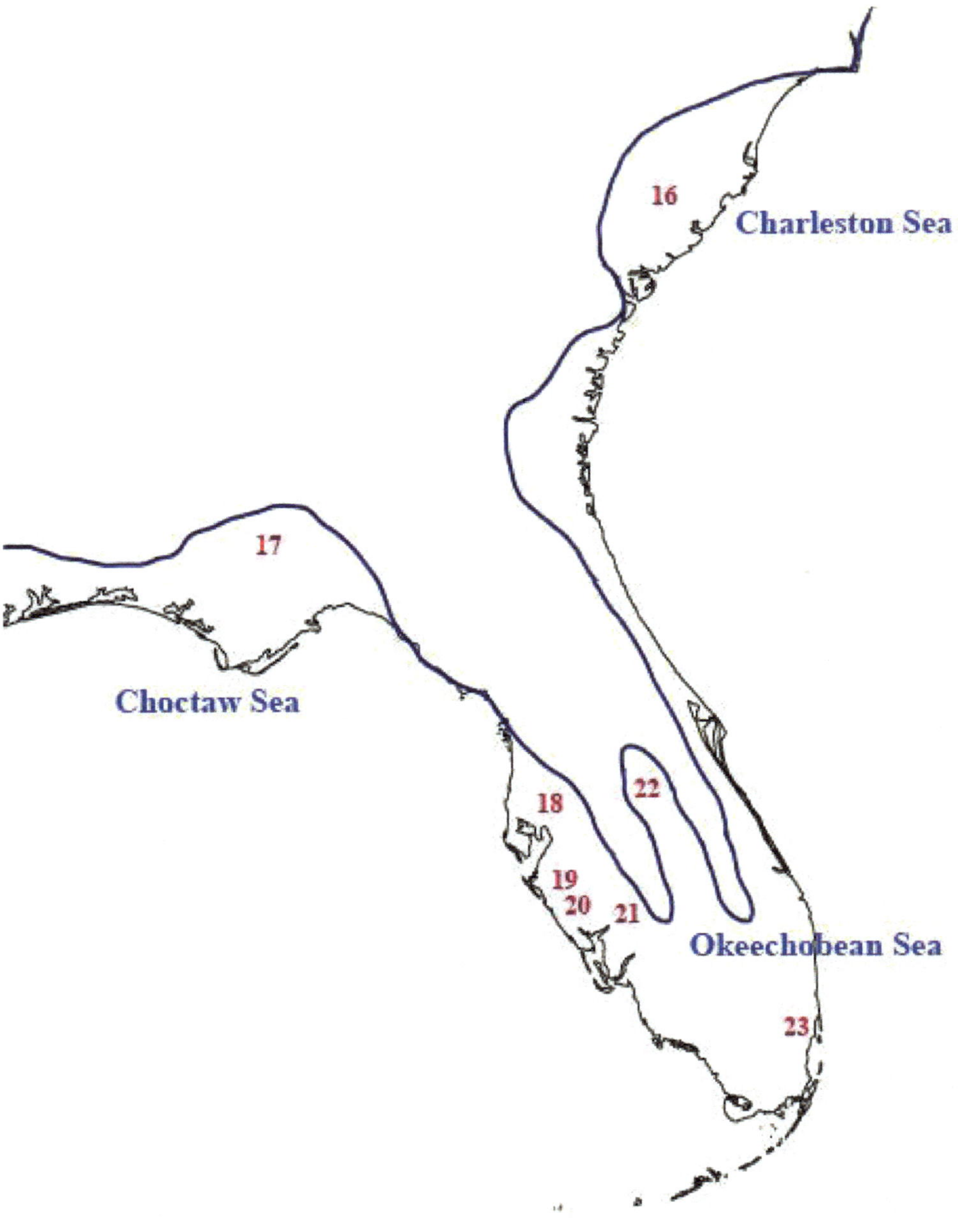

Map of the eastern United States showing some important southern localities where ecphoras have been regularly collected, including:

16= Eutawville, South Carolina (in quarries)
17= Apalachicola River, at Alum Bluff, Bristol, Florida (exposed in cliffs)
18= St. Petersburg and Tampa, Florida (blocks of limestone dredged from Tampa Bay)
19= Sarasota, Florida (in quarries)
20= Venice, Florida (in construction excavations)
21= Port Charlotte-Murdock area, Florida (in canal banks and housing excavations)
22= Kissimmee River, at Basinger (canal dredging and excavations)
23= Kendale Lakes, Miami, Florida (in lake excavations)

The known geographical range of the ecphoras extends from southern New Jersey in the north to Miami in the south, and as far west as the Apalachicola River in the Florida Panhandle. This area corresponds to the depositional centers of five paleoseas; the **Choctaw Sea** and **Okeechobean Sea** in Florida, the **Charleston Sea** in South Carolina and southern North Carolina, the **Albemarle Sea** in northern North Carolina and Virginia, and the **Salisbury Sea** in Maryland and southern New Jersey (see Petuch, 2004; Petuch and Drolshagen, 2010).

The Miocene Formations of Maryland and Their Ecphoras

Extending from Chesapeake Beach, northern Calvert County south to Drum Point, southern Calvert County, Maryland, the eroded coastal hills known as the "Calvert Cliffs" contain the best and most readily-accessible exposures of Miocene fossils found anywhere in North America (Figures 2.1 and 2.2). Forty species of ecphoras have been collected along the Calvert Cliffs and the adjacent Patuxent River, making this the single largest fauna of ecphoras known from one area. The geologic formations of the Chesapeake Miocene are shown on the following list, along with the members and their component Shattuck Zones and the subseas within which they were deposited. These units and their ecphora faunas were deposited in the Salisbury Sea's four subseas, the Old Church, the Calvert, the Patuxent, and the St. Mary's. The Miocene of southern New Jersey is also included here, as its single ecphora-bearing unit, the Shiloh Member of the Kirkwood Formation, was deposited along the Millville Delta in the northern Calvert Subsea during Shattuck Zone 10 (Plum Point Member) time (see Petuch, 2004: 101-110, figures 18 and 19). The Maryland and New Jersey Miocene includes the following geologic units:

OLD CHURCH FORMATION—Deposited within the Old Church Subsea, Salisbury Sea

Ecphorosycon tampaensis

CALVERT FORMATION—Deposited within the Calvert Subsea, Salisbury Sea
Fairhaven Member
Shattuck Zone 1--no ecphoras present
Shattuck Zone 2

Siphoecphora cf. *aurora* (only fragments)

Shattuck Zone 3—no ecphoras present
Plum Point Member
Shattuck Zone 4

Trisecphora (Eccentrella) chamnessi

Shattuck Zone 5—no ecphoras present
Shattuck Zone 6—no ecphoras present
Shattuck Zone 7—no ecphoras present
Shattuck Zone 8—no ecphoras present
Shattuck Zone 9—no ecphoras present
Shattuck Zone 10

Ecphorosycon pamlico
Chesathais whitfieldi
Chesathais ecclesiasticus
Trisecphora tricostata
Trisecphora (Recurvecphora) schmidti
Trisecphora (Eccentrella) eccentrica

Trisecphora (Matoaka) prunicola
Ecphora wardi

KIRKWOOD FORMATION (equivalent to Shattuck Zone 10, Calvert Formation)
Shiloh Member
Chesathais whitfieldi

CALVERT FORMATION—Deposited within the Calvert Subsea of the Salisbury Sea
Plum Point Member
Shattuck Zone 11
broken fragments only—*Ecphora (Powhatan)* cf. *calvertensis*
Shattuck Zone 12
Ecphorosycon kalyx
Trisecphora governorensis
Shattuck Zone 13—no ecphoras present
Calvert Beach Member
Shattuck Zone 14
Ecphora mattinglyi
Ecphora (Powhatan) calvertensis
Trisecphora scientistensis
Planecphora turneri
Trisecphora (Recurvecphora) martini
Shattuck Zone 15—no ecphoras present
Shattuck Zone 16
Trisecphora (Matoaka) bartoni
Ecphora (Powhatan) chesapeakensis

CHOPTANK FORMATION—Deposited within the Patuxent Subsea, Salisbury Sea
St. Leonard Member
Shattuck Zone 17
Ecphorosycon lindajoyceae
Chesathais donaldasheri
Trisecphora (Matoaka) smithae
Ecphora sandgatesensis
Ecphora (Rhombecphora) harasewychi
Planecphora vokesi
Drumcliff Member
Shattuck Zone 18
Chesathais lindae
Trisecphora (Matoaka) patuxentia
Ecphora meganae
Ecphora (Rhombecphora) rikeri

Planecphora choptankensis

Boston Cliffs Member

Shattuck Zone 19

Chesathais drumcliffensis

Trisecphora (Matoaka) shattucki

Ecphora amyae

Ecphora (Powhatan) williamsi

Planecphora delicata

ST. MARY'S FORMATION—Deposited within the St. Mary's Subsea, Salisbury Sea

Conoy Member

Shattuck Zone 20

Ecphora (Powhatan) conoyensis

Little Cove Point Member

Shattuck Zone 21

Ecphora (Powhatan) asheri

Shattuck Zone 22

Ecphora (Powhatan) germonae

Shattuck Zone 23—no ecphoras present

Windmill Point Member

Shattuck Zone 24

Ecphora (Powhatan) gardnerae

Chancellor Point Sandstone

"Shattuck Zone 25"

Ecphora (Powhatan) gardnerae angusticostata

Figure 2.1 View of a highly-fossiliferous exposure of the Drumcliff Member (Shattuck Zone 18) of the Choptank Formation at the base of a cliff near Matoaka Cottages, St. Leonards, Calvert County, Maryland, Chesapeake Bay. The contact with the overlying unfossiliferous clay unit can be seen to be very sharp and abrupt. Farther down the beach, the cliff has collapsed due to erosion and produced a large slump, which has buried the entire exposed section under layers of orange sand and clay.

Figure 2.2 Close-up view of an exposure of the Drumcliff Member (Shattuck Zone 18) of the Choptank Formation near Matoaka Cottages, St. Leonards, Calvert County, Maryland, showing a broken and partially-buried specimen of *Ecphora meganae* (center of the photo).

Eastover Formation	Cobham Bay Member	
	Claremont Manor Member	
St. Mary's Formation	Chancellor Point Sandstone	Zone 25
	Windmill Point Member	Zone 24
	Little Cove Point Member	Zone 23 Zone 22 Zone 21
	Conoy Member	Zone 20
Choptank Formation	Boston Cliffs Member	Zone 19
	Drumcliff Member	Zone 18
	St. Leonard Member	Zone 17
Calvert Formation	Calvert Beach Member	Zone 16 Zone 15 Zone 14
	Plum Point Member	Zone 13 Zone 12 Zone 11 Zone 10 Zone 9 Zone 8 Zone 7 Zone 6 Zone 5 Zone 4
	Fairhaven Member	Zone 3 Zone 2 Zone 1

Generalized Stratigraphic Column for the Chesapeake Bay Area of Maryland and Northern Virginia. Pink highlights indicate geological units that contain ecphora shells. Fairhaven Member is Burdigalian Age, Miocene. Plum Point Member and Calvert Beach Member are Langhian Age, Miocene. The Choptank Formation is Serravallian Age, Miocene. The St. Mary's Formation Conoy Member is Serravallian Age, Miocene, and Little Cove Point Member through Chancellor Point Sandstone are Tortonian Age, Miocene. The Eastover Formation Claremont Manor Member is Tortonian Age, Miocene, and the Cobham Bay Member is Messinian Age, Miocene. Zones 1-25 refer to the "Shattuck Zones" established by pioneer Maryland stratigrapher, George Shattuck (see Shattuck, 1904).

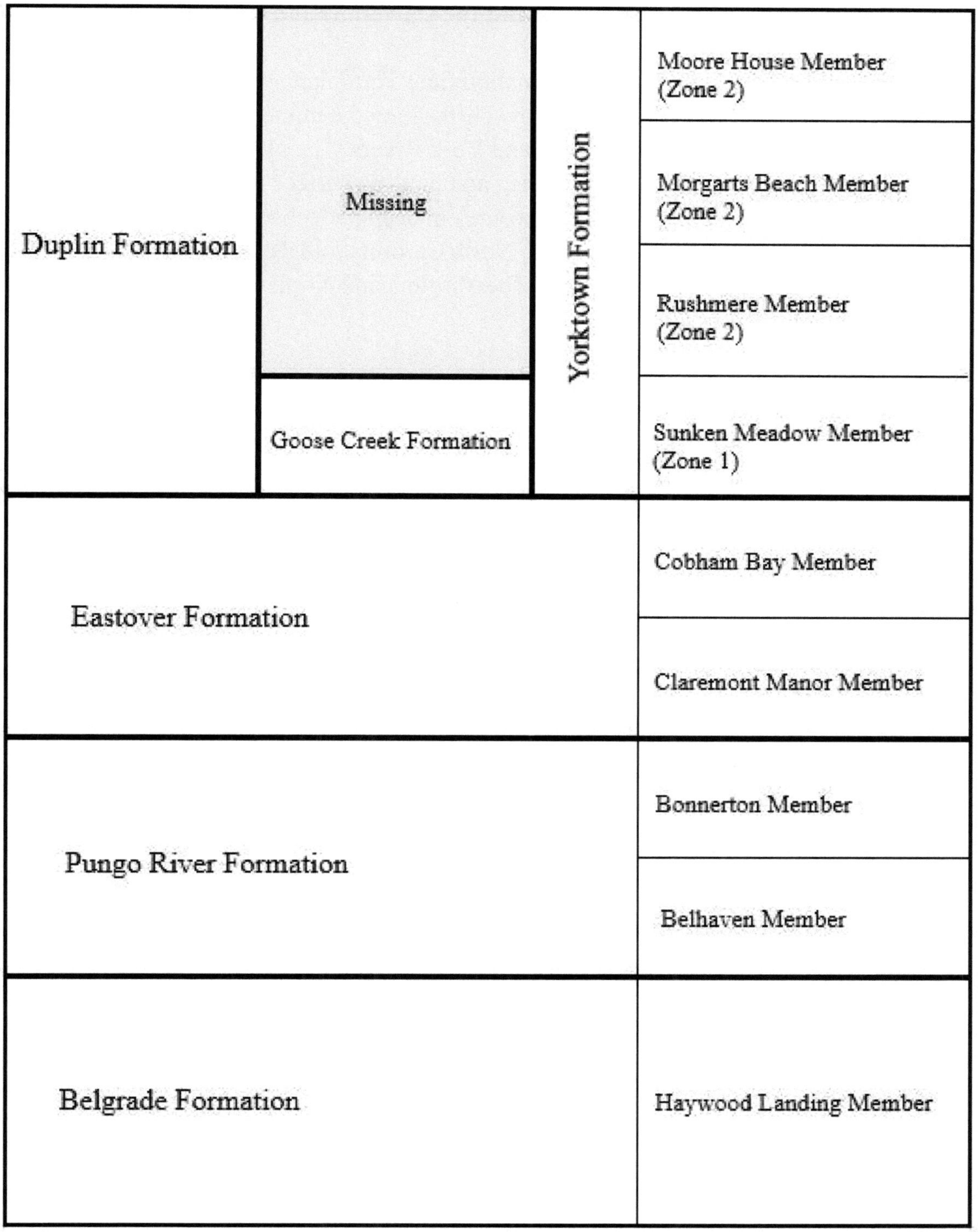

Generalized Stratigraphic Column for Virginia, North Carolina, and South Carolina. Every geological unit contains ecphora shells. The Belgrade Formation Haywood Landing Member is Aquitanian Age, Miocene. The Pungo River Formation Belhaven Member is Burdigalian Age, Miocene, and the Bonnerton Member is Langhian Age, Miocene. The Eastover Formation Claremont Manor Member is Tortonian Age, Miocene, and the Cobham Bay Member is Messinian Age, Miocene. The lower Duplin Formation / Goose Creek Formation / Sunken Meadow Member of the Yorktown Formation are Zanclean Age, Pliocene. The upper Duplin Formation / Yorktown Formation Rushmere Member through Moore House Member are Piacenzian Age, Pliocene.

The Miocene Formations of Virginia and the Carolinas and Their Ecphoras

Only the late Miocene Eastover Formation (late Tortonian and Messinian) is exposed at the surface in Virginia, primarily in low cliffs along the numerous tidewater rivers, such as the Rappahannock, Pamunkey, and York Rivers. The older Miocene formations, such as the Pungo River, are deeply buried and have no surface outcrops and can only be accessed through well corings or by exposures in deep phosphate mines, such as the PotashCorps Phosphate Mine in Aurora, North Carolina (old Texasgulf Lee Creek Phosphate Mine) (Figures 2.3 and 2.4). The Virginia and North Carolina Miocene includes the following geologic units:

BELGRADE FORMATION--Deposited within the Silverdale Subsea, Albemarle Sea
Haywood Landing Member
Ecphorosycon tampaensis
PUNGO RIVER FORMATION--Deposited within the Pamlico Subsea, Albemarle Sea
Belhaven Member
Siphoecphora aurora
Trisecphora precursor
Trisecphora (Eccentrella) chamnessi
Trisecphora (Recurvecphora) recurvicostata
Bonnerton Member
Ecphorosycon pamlico
Siphoecphora gerrardi
Chesathais pungoensis
Trisecphora tricostata
Trisecphora (Matoaka) carolinensis
Ecphora wardi
EASTOVER FORMATION--Deposited within the Rappahannock Subsea, Albemarle Sea
Claremont Manor Member
Ecphora (Powhatan) whiteoakensis
Cobham Bay Member
Ecphora (Powhatan) grulkei
Ecphora kochi

The Pliocene Formations of Virginia and the Carolinas and Their Ecphoras

Only a single Pliocene formation occurs in Virginia, the Yorktown and its four members. Like the underlying late Miocene Eastover Formation, the Pliocene Yorktown Formation crops out in low cliffs along several of the large tidewater rivers in southern Virginia, such as the York and the James Rivers. Farther south, the Yorktown equivalent

Duplin and Goose Creek Formations occur sporadically in small exposures in North and South Carolina, often in creeks and fill pits dug by local farmers (such as at Eagle Point). The largest and best exposure of the Goose Creek Formation ("Goose Creek Limestone") has recently been uncovered in the Martin Marietta Mine near Santee, South Carolina. Most of the ecphoras collected there were embedded in a soft, vuggy limestone at the base of the quarry excavation. The Virginia, North Carolina, and South Carolina Pliocene includes the following geologic units:

YORKTOWN FORMATION
Deposited within the Williamsburg Subsea, Albemarle Sea
Sunken Meadow Member (Yorktown Zone 1)
Ecphora leecreekensis
Ecphora (Powhatan) pachycostata

Figure 2.3 Close-up view of a block of phosphatic limestone from the Bonnerton Member of the Pungo River Formation, showing embedded specimens of *Trisecphora (Matoaka) carolinensis*. This fossiliferous block was excavated from the lowest beds of the PotashCorps Phosphate Mine (old Texasgulf Lee Creek Phosphate Mine), near Aurora, Beaufort County, North Carolina.

Figure 2.4 View of an excavation in the PotashCorp Phosphate Mine (formerly the Texasgulf Lee Creek Phosphate Mine) at Aurora, Beaufort County, North Carolina. Here, the crane is cutting into the dark phosphatic sands and clays of the Yorktown Formation (Zones 1 and 2), exposing richly-fossiliferous shell beds that contain seven different species of ecphoras.

Figure 2.5 Close-up view of an exposure of the Sunken Meadow Member (Zone 1) of the Yorktown Formation, showing a partially-buried *Ecphora (Powhatan) pachycostata* (to the right of the head of the geologic hammer and to the left of a fragment of whale rib). This richly-fossiliferous unit was excavated from the middle beds of the PotashCorps Phosphate Mine (old Texasgulf Lee Creek Phosphate Mine), near Aurora, Beaufort County, North Carolina.

Deposited within the Yorktown Subsea, Albemarle Sea
Rushmere Member (Yorktown Zone 2)
Ecphora quadricostata
Planecphora mansfieldi
Globecphora parvicostata
Morgarts Beach Member (Yorktown Zone 2)
Ecphora quadricostata umbilicata
Globecphora mardieae
Moore House Member (Yorktown Zone 2)
Ecphora quadricostata rachelae
DUPLIN FORMATION—Deposited within the Duplin Subsea, Charleston Sea
Ecphora quadricostata umbilicata
Ecphora quadricostata rachelae
Latecphora bradleyae
Planecphora mansfieldi
Globecphora parvicostata
GOOSE CREEK FORMATION—Deposited within the Santee Subsea, Charleston Sea
Ecphora quadricostata
Ecphora roxaneae dalli
Latecphora violetae
Globecphora parvicostata

The Miocene Formations of Southern Florida and Their Ecphoras

With no natural outcrops, the Miocene formations of southern Florida can only be accessed through canal excavations and housing construction projects. The Tampa Member of the Arcadia Formation, with its single ecphora, was originally exposed during dredging projects for deepening the ship channels in Tampa Bay. The boulders from these excavations were later used as ballast for ships and the locality became known as "Ballast Point". The Bayshore Formation is known only from canal digs near Port Charlotte in southwestern Florida and some exposures of the formation are still visible along several canal banks in the area (see Petuch and Berschauer, 2021 for details on the late Miocene geology and molluscan paleontology of southern Florida). The southern Florida Miocene includes the following geologic units:

ARCADIA FORMATION--Deposited within the Tampa Subsea, Okeechobean Sea
Tampa Member
Ecphorosycon tampaensis
BAYSHORE FORMATION--Deposited within the Charlotte Subsea, Okeechobean Sea
Cocoplum Member
Ecphora (Powhatan) whiteoakensis hopei

Port Charlotte Member
Globecphora floridana
Globecphora floridana streami

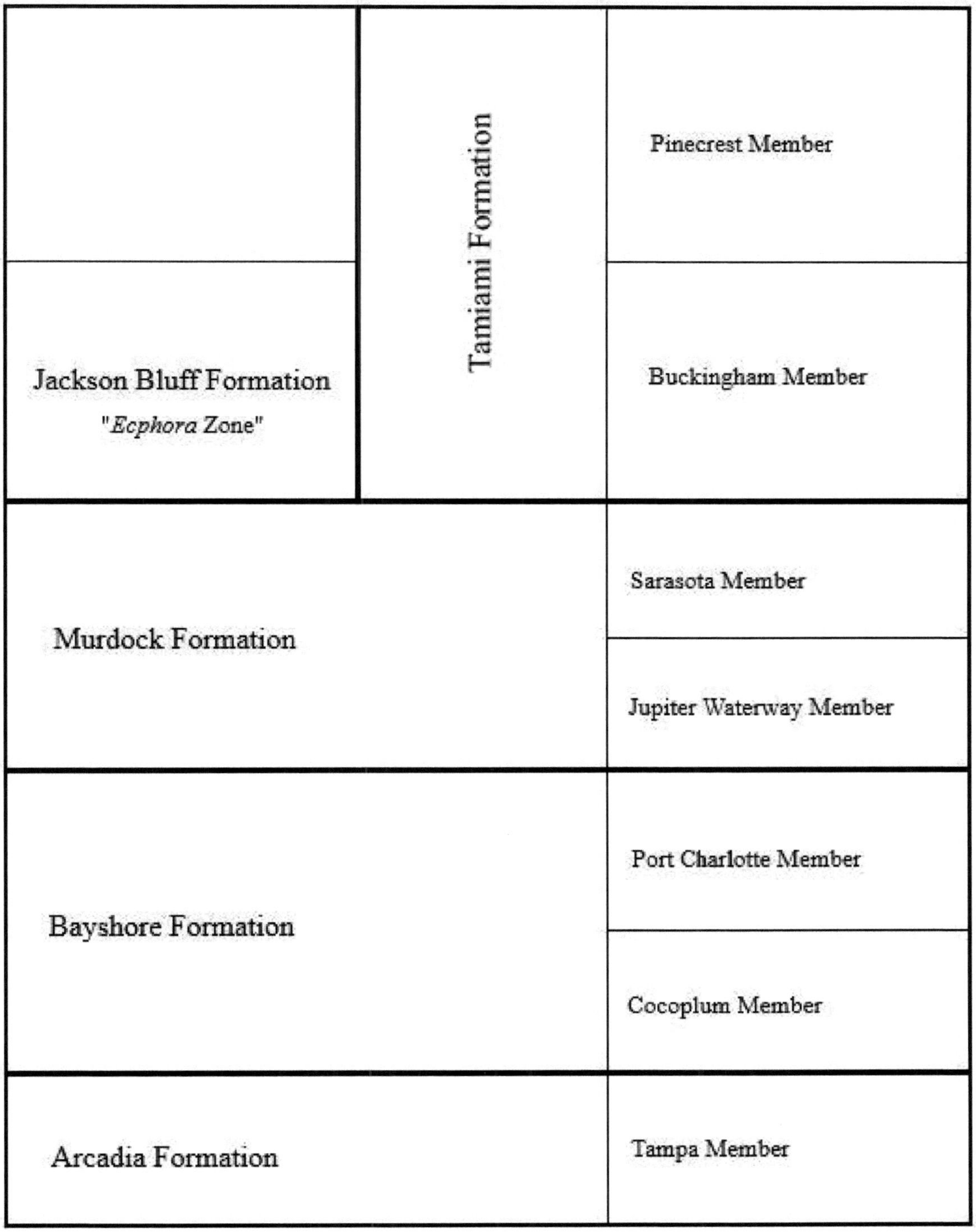

Generalized Stratigraphic Column for Northwestern Florida (Jackson Bluff Formation), and Southern Florida. Every geological unit contains ecphora shells. The Arcadia Formation is Aquitanian Age, Miocene. The Bayshore Formation is Messinian Age, Miocene. The Murdock Formation is Zanclean Age, Pliocene. The Jackson Bluff Formation "*Ecphora* Zone" Buckingham Member equivalent is early Piacenzian, Pliocene. The Tamiami Formation Buckingham Member is early Piacenzian, Pliocene, and the Pinecrest Member is late Piacenzian, Pliocene.

Figure 2.6 Close-up view of a highly fossiliferous exposure of the Pinecrest Member, Tamiami Formation in the APAC Pit at Sarasota, Sarasota County, Florida. A large, broken and barnacle-encrusted specimen of *Latecphora bradleyae* (aperture facing outward) can be seen at the very bottom of the photograph, directly below the large white *Triplofusus duplinensis* Horse Conch and to the left of the large *Carolinapecten* scallop.

The Pliocene Formations of Southern Florida and Their Ecphoras

Only two geological formations make up the southern Floridian Pliocene; the Zanclean-aged Murdock Formation (formerly Murdock Station Formation) and the Piacenzian-aged Tamiami Formation. Like the Miocene formations of southern Florida, there are no natural exposures of these formations and geologists and paleontologists are completely reliant on material brought up from quarries, canal excavations, and construction sites. Both formations were regularly exposed, for decades, in two large quarry complexes in Sarasota; the Ashland Petroleum and Asphalt Corporation (APAC) pits and the Quality Aggregates Inc. quarries. Large numbers of beautifully-preserved fossil mollusks were collected from these formations, particularly the Buckingham and Pinecrest Members of the Tamiami Formation (Figure 2.6). Another Tamiami Member, the Fruitville, lies on top of the Pinecrest, but since ecphoras became extinct at the end of Pinecrest time, this younger member is not included here (see Petuch and Berschauer, 2021 for detailed overviews of the mollusks of the Murdock and Tamiami Formations). The southern Floridian Pliocene includes the following geologic units:

MURDOCK FORMATION--Deposited within the Murdock Subsea, Okeechobean Sea

Jupiter Waterway Member

Ecphora quadricostata
Ecphora (Powhatan) pachycostata

Sarasota Member

Ecphora roxaneae
Planecphora hertweckorum

TAMIAMI FORMATION--Deposited within the Tamiami Subsea, Okeechobean Sea

Buckingham Member

Ecphora quadricostata striatula
Planecphora mansfieldi
Latecphora violetae

Pinecrest Member--Golden Gate Member (lower beds)

Ecphora quadricostata umbilicata
Ecphora quadricostata rachelae
Globecphora parvicostata
Latecphora bradleyae
Latecphora hertwecki

The Jackson Bluff Formation of Northwestern Florida and Its Ecphoras

Unlike southern Florida, the single Pliocene formation of northwestern Florida ("Florida Panhandle"), the Jackson Bluff, has large exposures that are readily available for both study and collecting. Located at the base of the high Alum Bluff along the

Apalachicola River in Liberty County, Florida (Figure 2.7), the lower fossiliferous bed (the "*Ecphora* Zone") crops out all along the edge of the river. The northwestern Floridian Pliocene includes the following geologic unit:

JACKSON BLUFF FORMATION--Deposited within the Jackson Subsea, Choctaw Sea
Lower Bed (The "*Ecphora* Zone")
Ecphora quadricostata striatula
Planecphora mansfieldi

The following chapters of this book are arranged evolutionarily; with the oldest genus (*Ecphorosycon*) being described and discussed in Chapter 3 and the youngest genus (*Latecphora*) being described and discussed in Chapter 8. Within each chapter, the taxa are arranged geochronologically, with the oldest species being listed first and the youngest species being listed last.

Figure 2.7 View of Alum Bluff along the Apalachicola River near Bristol, Liberty County, Florida, showing exposures of the Jackson Bluff Formation. The dark-colored layer at the base of the cliff is an undescribed member of the Jackson Bluff that is referred to as the "*Ecphora* Zone", where two species of ecphoras are frequently collected.

CHAPTER 3. The Siphonate and Rapaniform Ecphoras

Three generic groups of ecphoras look quite different from the standard three and four-ribbed shells that characterize this group of muricids. These genera, *Ecphorosycon*, *Siphoecphora*, and *Chesathais*, differ from the other well-known groups by having multiple cordlike ribs, most of which are adherent to the body whorl, causing the shells to resemble living muricids in the subfamily Rapaninae. Of the three rapaniform genera, *Ecphorosycon* is the most primitive and is the first to appear in the fossil record as the small species, *Ecphorosycon tampaensis*, from the Aquitanian Miocene Tampa Member of the Arcadia Formation and the Haywood Landing Member of the Belgrade Formation. This elongated, generalized ecphora genus is the stem-stock for all subsequent ecphora genera and species radiations. By the Burdigalian Miocene, the long-siphoned genus *Siphoecphora* had evolved and co-existed with descendants of the Aquitanian *Ecphorosycon tampaensis* in North Carolina and Maryland. At the same time, the multi-ribbed rapaniform ecphora group, *Chesathais*, evolved and lived together with *Ecphorosycon* and *Siphoecphora* species along the North Carolina coast. By the Langhian Miocene, the long-siphoned, buyscon-like genus *Siphoecphora* became extinct, but the other two archaic genera survived until the end of the Serravallian Miocene.

When describing two unusual new ecphoras found in the Pungo River Formation of North Carolina (*pamlico* and *aurora*), Druid Wilson (1987) placed both odd-looking species in the genus *Stenomphalus* Sandberger, 1861. That taxon, however, refers to a group of tiny species in the subfamily Rapaninae, averaging only 15 mm in length, and confined to the Oligocene and Miocene formations of Europe (oddly, Wilson gave no size measurements for the tiny, *Stenomphalus* species that he illustrated, creating the false impression that they were much larger shells than they actually are). The American species named by Wilson, on the other hand, are very large shells that frequently reach 80 mm in length, possess all the shell characteristics of the Ecphorini, and are clearly not members of the Rapaninae. Although illustrating three species of European *Stenomphalus*, Wilson never gave their sizes nor discussed the size discrepancies seen between the North American and European taxa. The shell sculpture similarities described by Wilson seem to represent examples of convergent evolution and do not represent any direct evolutionary relationship. Wilson's Pungo River species clearly did not belong in *Stenomphalus* and were later recognized as representing two previously-unnamed ecphora groups. These were described as the new genera *Ecphorosycon* and *Siphoecphora* by the senior author in 1988. Both groups are discussed in the following sections of this chapter, along with the rapaniform genus *Chesathais*.

The Genus *Ecphorosycon* Petuch, 1988

SYSTEMATICS
Class Gastropoda Cuvier, 1797
Subclass Orthogastropoda Ponder and Lindberg, 1995
Superorder Caenogastropoda Cox, 1960
Order Sorbeoconcha Ponder and Lindberg, 1997
Infraorder Neogastropoda Wenz, 1938
Superfamily Muricoidea Rafinesque, 1815
Family Muricidae Rafinesque, 1815
Subfamily Ocenebrinae Cossmann, 1903
Tribe Ecphorini Petuch and Berschauer, new tribe
Genus *Ecphorosycon* Petuch, 1988

Diagnosis: Shells large for the Tribe Ecphorini, inflated, pyriform or cylindrical in outline; spires low; shoulders sloping, with flat or slightly-angled subsutural area; shells ornamented with three or four main ribs and numerous smaller riblets and threads between ribs; main ribs thin and bladelike on some species, rounded on others; siphoal canals well developed, elongated, heavily sculptured with ribs and small cords; umbilici well developed, flaring in some species.
Type Species: *Ecphorosycon pamlico* (Wilson, 1987), Pungo River Formation of North Carolina and Calvert Formation of Maryland, Langhian Age, Miocene.
Species in *Ecphorosycon*: *Ecphorosycon tampaensis* (Dall, 1890), Tampa Member, Arcadia Formation; *Ecphorosycon pamlico* (Wilson, 1987), Belhaven Member, Pungo River Formation and the Plum Point Member (Shattuck Zone 10), Calvert Formation; *Ecphorosycon kalyx* (Petuch, 1988), Plum Point Member (Shattuck Zone 12, Calvert Formation); *Ecphorosycon lindajoyceae* Petuch, 1993, St. Leonard Member (Shattuck Zone 17, Choptank Formation).
Discussion: This primitive group of ecphoras first appears in the earliest Aquitanian Miocene (Oligocene-Miocene boundary) as the widespread species *Ecphorosycon tampaensis*, which ranged from Maryland to South Carolina, and south to southern Florida. By the early Langhian Miocene, the members of the genus had evolved larger shells with more angled shoulders, and were actually the largest muricid gastropods in eastern North America at that time. *Ecphorosycon* survived into the beginning of the Serravallian Age as the very large and heavy *E. lindajoyceae*, but disappeared from the fossil record immediately afterward. The genus may have preferred warmer water conditions and the abrupt cooling of the oceans in the early Serravallian (caused by the Ries-Steinheim asteroid impact; see Chapter 1) appears to have killed-off the last remaining *Ecphorosycon* populations in the Salisbury Sea.

Ecphorosycon tampaensis (Dall, 1890)
(Figure 3.1 A, B)

Description: (Reprinted from the original description, as *Rapana tampaensis*; Dall, 1890: 153) "Shell rather small, short-spired; last whorl much the largest; spiral sculpture of eight or nine primary ridges, elevated and square-sided, of which one is near the suture, two others (a little larger) with subequal interspaces in front of the first, then an interspace of greater width with the strongest spiral in front of it, then the peripheral interspace, twice as wide as the strongest spiral, then another strong spiral followed by a narrower channel, and three more basal spirals with diminishing interspaces on the base, and three or four rather obscure and less elevated spiral cords on the canal; of secondary spirals much smaller than the primaries, there are one or two in the wide peripheral channel, and sometimes a single one elsewhere; lastly, the whole surface is more or less sculptured by fine, incised spiral lines. The transverse sculpture is of fine wrinkles, in harmony with the incremental lines, which cover most of the shell; the ribs are slightly but irregularly undulated in some places; aperture ovate, pillar-lip with a thin, smooth callus; edge of the outer lip undulated by the sculpture, the interior is lirate, with ten sharp, prominent lirae disposed somewhat in pairs; canal very narrow and deep, shorter than the aperture, umbilicus flaring, variable in diameter, deep, bounded by the rounded edge of the fasciole, within axially striated, but not otherwise sculptured."
Type Specimen: HOLOTYPE: length 35 mm, from the Tampa Member of the Arcadia Formation at Ballast Point, Pinellas County, Tampa Bay, Florida. In the type collection of the Department of Paleobiology, National Museum of Natural History, Smithsonian Institution, USNM 165096.
Type Locality: (From Dall, 1915: 78) "Tampa silex beds at Ballast Point, Tampa Bay, Florida."
Stratigraphic Range: Restricted to the early Aquitanian Age of the Early Miocene and found in the Tampa Member of the Arcadia Formation in Florida, the Belgrade Formation of North Carolina, and the Old Church Formation of Maryland. Originally thought to be late Oligocene in age, but now known to date from the Aquitanian Age of the southeastern United States.
Etymology: Named for Tampa, Hillsborough County, Florida and Tampa Bay.
Discussion: Along with being the oldest-known true ecphora, *Ecphorosycon tampaensis* is also one of the most wide-ranging species of its entire Tribe; being found in the fossil beds of the Old Church Formation of Maryland, the "Silverdale Beds" of the Belgrade Formation of South Carolina, and the Tampa Member of the Arcadia Formation of southern Florida.

Ecphorosycon pamlico (Wilson, 1987)
(Figure 3.2 A, B)

Description: (Reprinted from the original description; as *Ecphora (Ecphora) pamlico*; Wilson, 1987: 24) "Shells moderately large with angle-sided whorls; three prominent spiral thin flange-like costae at the whorl angles. Secondary sculpture of many spiral bands between costae; bands becoming obsolete or nearly so on the flattened shoulder; posterior interspace between the costae wider than anterior interspace. Aperture ovate; anteriorly produced into a very narrow, broadly curved canal. Umbilicus large, rounded and bordered by a regularly and weakly stepped (or serrated) carina."

Type Specimens: HOLOTYPE: length 65 mm, width 65 mm, from the basal beds in the PotashCorps Phosphate Mine, Aurora, Beaufort County, North Carolina (old Lee Creek Phosphate Mine), USNM 647668; PARATYPES: length 58 mm, from the same locality and formation as the holotype, USNM 647671.

Type Locality: Found in the Bonnerton Member of the Pungo River Formation, basal beds of the PotashCorps Phosphate Mine, Aurora, Beaufort County, North Carolina.

Stratigraphic Range: Confined to the early Langhian Age formations of the early Miocene of the eastern United States, from Delaware to at least North Carolina. Found in the Bonnerton Member of the Pungo River Formation of North Carolina, the Plum Point Member (Shattuck Zone 10) of the Calvert Formation of Maryland, and at the Pollack Farm site, Kent County, Delaware (Ward, 1998; misidentified as "*Ecphora tricostata*").

Etymology: Named for the Pamlico River, which is adjacent to the PotashCorps Phosphate Mine (the old Lee Creek Phosphate Mine).

Discussion: Martin (1904; plate 52, figure 8) illustrates a specimen of *Ecphorosycon pamlico* from the Calvert Formation at Plum Point, Maryland, but identified it only as an "aberrant form" of *Trisecphora tricostata* (see Chapter 4); it took 83 years for *E. pamlico* to be described, from when it was first illustrated by Martin (from a Maryland specimen) to when it was recognized as new and described by Wilson (from a North Carolina specimen). Subsequently, other closely-related species were discovered and described, demonstrating that the genus *Ecphorosycon* represents a distinct evolutionary clade with a larger stratigraphic range than previously thought. The shell was collected at Plum Point, the same locality as the specimen illustrated here on Figure 3.2.

Ecphorosycon kalyx (Petuch, 1988)
(Figure 3.3 A, B, C)

Description: (Reprinted from the original description; as *Ecphora kalyx*; Petuch, 1988: 14-15) "Shell small for genus, inflated, with sharp-angled shoulder; body whorl with 5 well-developed ribs; rib along shoulder expanded posteriorward, producing canaliculated, cup-like spire whorls; anteriormost 2 ribs smaller and less developed than other ribs; areas between ribs heavily sculptured with raised threads and fine spiral cords; siphonal canal well developed, heavily sculptured with spiral cords."
Type Specimens: HOLOTYPE: length 17 mm (incomplete), width 16 mm, upper bed at Plum Point, Calvert Cliffs, Calvert County, Maryland, Shattuck Zone 12 of the Plum Point Member, Calvert Formation, USNM 424258; PARATYPES: lengths 29 and 68 mm, from Shattuck Zone 12 of the Plum Point Member, Calvert Formation, north of Scientist's Cliffs, Calvert Cliffs, Calvert County, Maryland, in the Petuch research collection.
Type Locality: In the upper bed (Shattuck Zone 12) at Plum Point, Calvert Cliffs, Calvert County, Maryland, Plum Point Member of the Calvert Formation.
Stratigraphic Range: Confined to Shattuck Zone 12 of the Plum Point Member, Calvert Formation; late Langhian Age Miocene.
Etymology: From the Greek word, "kalyx" (a drinking vessel), in reference to the characteristic sunken, cuplike spire whorls.
Discussion: *Ecphorosycon kalyx* is similar to the older and stratigraphically-lower *E. pamlico*, but differs in having much stronger, wider, and thicker ribs on the body whorl. The spire whorls are also much more canaliculated than those of the older, ancestral species. Although the holotype is actually a broken juvenile specimen, it still retains the key morphological characteristics seen on adult shells.

Ecphorosycon lindajoyceae Petuch, 1993
(Figure 3.4 A, B, C)

Description: (Reprinted from the original description; Petuch, 1993: 164-165, figures 5, 9) "Shell inflated, with globose body whorl and distinctly pyriform outline; spire protracted, scalariform; shoulder sloping; subsutural area flattened, producing stepped spire whorls; body whorl ornamented with 4 rounded, thin, low, adherent cords; cord along shoulder of body whorl projecting upward (posteriorward), becoming progressively more bladelike on earlier whorls; entire body whorl, spire whorls, and siphonal canal sculptured with spiral threads; spiral threads give entire shell silky texture; spire whorls ornamented with 2 thin cords, with cord along shoulder being bladelike in form; siphonal canal proportionally short and broad for genus; umbilicus wide, flaring, well developed."
Type Specimens: HOLOTYPE: length 55 mm (incomplete), in Shattuck Zone 17, St. Leonard Member of the Choptank Formation at Sandgates, St. Mary's County, Maryland; at the waterline of the Patuxent River), UF 21466; PARATYPE: length 69 mm (crushed and flattened, partially reconstructed) , same locality and formation as the holotype, in the Petuch research collection.
Type Locality: Along the waterline of the Patuxent River at Sandgates, St. Mary's County, Maryland, in Shattuck Zone 17 (St. Leonard Member) of the Choptank Formation.
Stratigraphic Range: Confined to the St. Leonard Member of the Choptank Formation, middle Serravallian Age of the middle Miocene. To date, collected only in the area of Sandgates, St. Mary's County, along a small stretch of the Patuxent River.
Etymology: Named Linda Joyce Petuch, wife of the senior author, who aided in the collection of fossil mollusks along the St. Mary's River and the Calvert Cliffs.
Discussion: The Choptank Formation *Ecphorosycon lindajoyceae* is most similar to *E. pamlico* from Zone 10 (Plum Point Member) of the Calvert Formation of Maryland and from the Bonnerton Member of the Pungo River Formation of North Carolina. The new species differs from *E. pamlico* in being a much wider, much more inflated shell with a wider, less elongated siphonal canal. The spire of *E. pamlico* is also higher and much more protracted than the spire of *E. lindajoyceae*, which is distinctly lower and more depressed. The spiral threads on the spire, body whorl, and siphonal canal of *E. lindajoyceae* are also much finer than those of the coarsely-sculptured *E. pamlico*. This is the last species of *Ecphorosycon* found in the fossil record and the genus became extinct after Choptank time (middle Serravallian Age of the middle Miocene).

Iconography of *Ecphorosycon* Species

Figure 3.1 ***Ecphorosycon tampaensis*** **(Dall, 1890)**
A, B= typical specimen, length 21 mm. Collected at Ballast Point, Pinellas County, Florida. An agatized pseudomorph from the Tampa Member, Arcadia Formation. Also found in the Belgrade Formation of North Carolina and the Old Church Formation of Maryland.

Figure 3.2 ***Ecphorosycon pamlico*** **(Wilson, 1987)**
A, B= typical specimen, length 74 mm. Collected at Plum Point, Calvert County, Maryland; from the Plum Point Member (Shattuck Zone 10), Calvert Formation. Originally described from the Belhaven Member of the Pungo River Formation of North Carolina. This is the type species of the genus *Ecphorosycon*.

Figure 3.3 ***Ecphorosycon kalyx*** **(Petuch, 1988)**
A, B= paratype specimen, length 68 mm. **C**= juvenile paratype, length 29 mm. Both specimens were collected in Shattuck Zone 12, north of Scientist's Cliffs, Calvert County, Maryland; in the Plum Point Member, Calvert Formation.

Figure 3.4 ***Ecphorosycon lindajoyceae*** **Petuch, 1993**
A, B= slightly crushed and flattened paratype, length 69 mm; **C=** juvenile specimen, length 22 mm. Both specimens were collected in Shattuck Zone 17 at Sandgates, St. Mary's County, along the Patuxent River, Maryland; in the St. Leonard Member, Choptank Formation. This is the last member of the genus *Ecphorosycon* and the genus became extinct by the end of Choptank time.

The Genus *Siphoecphora* Petuch, 1988

With their proportionally-long and extended siphonal canals, the longest of any ecphora genus or species, members of this genus somewhat resemble ornate versions of busycon whelks (Family Busyconidae). With their three broad ribs, especially prominent on juvenile shells, *Siphoecphora* may represent the ancestral stock from which the three-ribbed genus *Trisephora*, and its subgenera, evolved. The genus is known only from beds of late Burdigalian and Langhian age and became extinct before the Serravallian Age.

SYSTEMATICS
Class Gastropoda Cuvier, 1797
Subclass Orthogastropoda Ponder and Lindberg, 1995
Superorder Caenogastropoda Cox, 1960
Order Sorbeoconcha Ponder and Lindberg, 1997
Infraorder Neogastropoda Wenz, 1938
Superfamily Muricoidea Rafinesque, 1815
Family Muricidae Rafinesque, 1815
Subfamily Ocenebrinae Cossmann, 1903
Tribe Ecphorini Petuch and Berschauer, new tribe
Genus *Siphoecphora* Petuch, 1988

Diagnosis: Shells pyriform, with low spires rounded shoulders and body whorls, and long well developed siphonal canals; body whorls sculpted with three very wide, flattened, and only slightly raised bandlike ribs; ribs heavily sculpted with three deeply-incised grooves and smaller incised threads; wide ribs separated by wide, shallow grooves; grooves ornamented with fine, raised threads; siphonal canals heavily sculptured with coarse raised threads and cords; apertures wide, rounded, flaring; umbilici narrow; spire whorls canaliculated.
Type Species: *Siphoecphora aurora* (Wilson, 1987), lower beds (Belhaven Member) of the Pungo River Formation, early Langhian Miocene of North Carolina.
Species in *Siphoecphora*: *Siphoecphora aurora* (Wilson, 1987), Belhaven Member, Pungo River Formation; *Siphoecphora gerrardi* Petuch and Berschauer, n. sp., Bonnerton Member, Pungo River, Formation.

Siphoecphora aurora (Wilson, 1987)
(Figure 3.5 A, B, C, D)

Description: (Reprinted from the original description, as *Ecphora (Stenomphalus) aurora*; Wilson, 1987: 24-25) "Shell moderately large for genus, pyriform with rounded whorls bearing three primary low but prominent spiral bands, which in turn bear secondary bifid spiral bands; the anterior primary band hardly set off from the anterior secondary bifid spiral bands of the anterior of the body whorl; shoulder flattened and sunken below edge of posterior primary band. Broadly ovate aperture anteriorly produced into a slightly curved siphonal canal. Umbilicus anteriorly produced and bordered by a rudely produced stepped carina."

Type Specimens: HOLOTYPE: length 80 mm, width 55 mm, from the basal beds in the PotashCorps Phosphate Mine, Aurora, Beaufort County, North Carolina (old Lee Creek Phosphate Mine), USNM 647654; PARATYPES: 2 specimens, lengths 30.5 mm and 44.5 mm, from the same locality and formation as the holotype, USNM 647655 and USNM 647656.

Type Locality: Found in the Belhaven Member of the Pungo River Formation, basal beds of the PotashCorps Phosphate Mine, Aurora, Beaufort County, North Carolina.

Stratigraphic Range: Confined to the latest Burdigalian Age of the early Miocene of North Carolina, in the Belhaven Member of the Pungo River Formation. May also be present in Shattuck Zone 2, Fairhaven Member of the Calvert Formation of Maryland, but is only known from broken fragments that were found imbedded in an indurated sandstone ledge.

Etymology: Named for Aurora, Beaufort County, North Carolina, the type locality.

Discussion: As shown here below, young specimens of *Siphoecphora aurora* resemble *Trisecphora* species in having three distinct, broad, raised ribs around the body whorl. These broad ribs eventually flatten out and become adherent to the body whorl as the shell matures, producing the characteristic corded appearance that is seen here on Figure 3.5A, B.

Siphoecphora aurora - juvenile specimen 49 mm in length.

Siphoecphora gerrardi Petuch and Berschauer, new species
(Figure 3,6 A, B, C, D)

Description: Shell moderately large for genus, pyriform with rounded whorls bearing three primary low but prominent spiral bands, which in turn bear secondary bifid spiral bands; the anterior primary band hardly set off from the anterior secondary bifid spiral bands of the anterior of the body whorl; shoulder flattened and sunken below edge of posterior primary band. Broadly ovate aperture anteriorly produced into a slightly curved siphonal canal. Umbilicus anteriorly produced and bordered by a rudely produced stepped carina.
Type Specimens: HOLOTYPE: length 31 mm, width 23.8 mm, Bonnerton Member, Pungo River Formation at Aurora, North Carolina, PRI 104772; PARATYPE: length 26 mm, from the same locality and formation as the holotype, in the Petuch research collection.
Type Locality: Embedded in phosphatic limestone in the lower beds of the Lee Creek Phosphate Mine (now PotashCorps Phosphate Mine), Bonnerton Member, Pungo River Formation.
Stratigraphic Range: Confined to the Bonnerton Member of the Pungo River Formation, Langhian Age of the early middle Miocene.
Etymology: Named for Dr. Edward Gerrard of Hickory, North Carolina, who assisted the senior author while conducting field work in the Pungo River Formation exposures at Aurora.
Discussion: The Bonnerton Member *Siphoecphora gerrardi* is similar to the older and stratigraphically-lower *S. aurora* from the Belhaven Member, but differs in being a smaller, more slender shell with distinctly smaller and thinner ribs on the body whorl. These ribs do not flatten and become adherent when the shell matures, as in *S. aurora*, but remain raised above the body whorl into adulthood.

Iconography of *Siphoecphora* Species

Figure 3.5 ***Siphoecphora aurora*** **(Wilson, 1987)**
A, B= typical adult specimen, length 70 mm; found embedded in slabs of phosphatic limestone from the basal beds of the PotashCorps Phosphate Mine, Aurora, Beaufort County, North Carolina; in the Belhaven Member, Pungo River Formation. Also known as fragments from Shattuck Zone 2 of the Fairhaven Member, Calvert Formation, Calvert County, Maryland. This is the type species of the genus *Siphoecphora*.

Figure 3.6 ***Siphoecphora gerrardi*** **Petuch and Berschauer, new species**
A, B= holotype, length 31 mm. **C, D=** paratype, length 26 mm. Both specimens were collected in the PotashCorps Mine, Aurora, Beaufort County, North Carolina; from the Bonnerton Member of the Pungo River Formation.

The Genus *Chesathais* Petuch, 1988

In general shell shape, form, and size, members of *Chesathais* resemble rapanine muricids of the heavily-ribbed, Australian-New Zealand genus *Dicathais* more than they do other known Eastern American Miocene ocenebrines. The raised cords inside the aperture of *Chesathais* also set it apart from any other known group of American Miocene ocenebrine muricids. Species in *Chesathais*, with their multiple low peripheral cords, also resemble members of the related genera *Siphoecphora* and *Ecphorosycon*, but differ in having more fusiform shells with much shorter and less-developed siphonal canals. The known species of *Chesathais* are listed in the following section.

SYSTEMATICS
Class Gastropoda Cuvier, 1797
Subclass Orthogastropoda Ponder and Lindberg, 1995
Superorder Caenogastropoda Cox, 1960
Order Sorbeoconcha Ponder and Lindberg, 1995
Infraorder Neogastropoda Wenz, 1938
Superfamily Muricoidea Rafineasque, 1815
Family Muricidae Rafinesque, 1815
Subfamily Ocenebrinae Cossmann, 1903
Tribe Ecphorini Petuch and Berschauer, new tribe
Genus *Chesathais* Petuch, 1988

Diagnosis: Shells average size for subfamily, high-spired and fusiform in shape; ribs low and cordlike, adherent; shells with 4 or 5 primary ribs and 3-5 secondary ribs between primaries; areas between ribs slightly fimbriated, especially in juvenile specimens; umbilici poorly-developed, nonexistent in many specimens; siphonal canal slightly developed, heavily sculptured with low raised cords; interior of aperture with 8-12 large raised cords on aragonitic wall; corded apertural wall often missing in worn, eroded specimens.
Type Species: *Chesathais lindae* Petuch, 1988, Shattuck Zone 18 of the Drumcliff Member, Choptank Formation.
Species in *Chesathais*: *Chesathais pungoensis* Petuch and Berschauer, n.sp., Belhaven Member, Pungo River Formation; *Chesathais whitfieldi* Petuch, 1988, Kirkwood Formation and Plum Point Member (Shattuck Zone 10), Calvert Formation; *Chesathais ecclesiasticus* (Dall, 1915), Plum Point Member (Shattuck Zone 10), Calvert Formation; *Chesathais donaldasheri* Petuch, 1988, St. Leonard Member (Shattuck Zone 17), Choptank Formation; *Chesathais lindae* Petuch, 1988, Drumcliff Member (Shattuck Zone 18), Choptank Formation; *Chesathais drumcliffensis* Petuch, 1988, Boston Cliffs Member (Shattuck Zone 19), Choptank Formation.
Etymology: Named as a combination of "Chesapeake", for Chesapeake Bay, and "*Thais*", in recognition of the members of this ocenebrine group's resemblance to rapanine muricids in the genus *Thais*.

Chesathais pungoensis Petuch and Berschauer, new species
(Figure 3.7 A, B)

Description: General shell form and structure as for genus; shell encircled with four thick, rounded primary ribs and numerous thin, smaller cords between main ribs; shoulder rib slightly raised, extended posteriorly, producing shallowly canaliculated spire whorls; siphonal canal proportionally large, well developed, heavily sculpted with strong raised threads and small cords
Type Specimen: HOLOTYPE: Length 21 mm (fragmentary), width 16.2 mm, Belhaven Member, Pungo River Formation at Aurora, North Carolina, PRI 104773.
Type Locality: From the lowest beds in the Lee Creek Phosphate Mine (now PotashCorps Phosphate Mine), Aurora, Beaufort County, North Carolina.
Stratigraphic Range: Confined to the Belhaven Member of the Pungo River Formation of North Carolina, early Langhian Age of the early middle Miocene.
Etymology: Named for the Pungo River Formation of North Carolina.
Discussion: *Chesathais pungoensis* is most similar to the younger and stratigraphically-higher *C. whitfieldi* from the Kirkwood and Calvert Formations, but differs in having fewer primary ribs on the body whorl and in having much stronger spiral sculpture between the individual cords. The main ribs are also proportionally larger, wider, and thicker than those seen on the younger descendant species.

Chesathais whitfieldi Petuch, 1988
(Figure 3.8 A, B)

Description: (Reprinted from the original description; Petuch, 1988: 113) "Shell fusiform, biconic, with high, elevated spire; body whorl and spire sculptured with low, thin, cords, numbering 12 on the body whorl; areas between main cords sculptured with fine spiral threads; siphonal canal short, stubby."

Type Specimen: HOLOTYPE: Juvenile, length 15 mm, from an exposure of the Kirkwood Formation in a quarry at Jericho, Cumberland County, New Jersey (originally illustrated in Whitfield, 1894: 100, plate 17 figures 9 and 10 and referred to as "*Lyrosoma sulcosa*"); in the Whitfield New Jersey Collection in the Department of Paleobiology, National Museum of Natural History, Smithsonian Institution. The large adult specimen shown here on Figure 3.8 A, B could be designated to lectotype if the juvenile holotype cannot be found in the Smithsonian collections.

Type Locality: In micaceous marls in a quarry at Jericho, Cumberland County, New Jersey, Kirkwood Formation.

Stratigraphic Range: Confined to the Kirkwood Formation of New Jersey and the contemporaneous Plum Point Member (Shattuck Zone 10) of the Calvert Formation of Maryland; both Langhian Age of the early middle Miocene.

Etymology: The taxon honors Robert P. Whitfield, pioneer New Jersey paleomalacologist, who illustrated this new species in his 1894 monograph.

Discussion: Although originally thought to be restricted to the Kirkwood Formation (Petuch, 1988:113), *Chesathais whitfieldi* is now known from the Plum Point Member of the Calvert Formation, with several adult specimens having been collected in Shattuck Zone 10 at Plum Point, Calvert County, Maryland. These fully-grown adult specimens demonstrate that the 12 strong cords seen on juvenile specimens become reduced in number as the animal matured, with only 5 main cords being present on larger specimens (seen here on Figure 3.8A, B). The fine threads between the main cords on juvenile specimens are retained into adulthood. A faint sixth cord is present around the posterior end of the siphonal canal on many specimens (like the one shown here).

Chesathais ecclesiasticus (Dall, 1915)
(Figure 3.9 A, B)

Description: Dall (1915:78) never formally described this Maryland distinctive species, instead simply referring to it as *Rapana ecclesiasticus* and citing and re-illustrating the specimen shown in Martin, 1904 (plate 52, figure 9). The species was redescribed by the senior author (Petuch, 1988: 112-113) and this is reprinted here: "Shell elongated, fusiform, somewhat biconical; shoulder rounded; subsutural area planar, producing narrow shelf; spire whorls turriculate, scalariform; umbilicus narrow, poorly developed; body whorl ornamented with seven to ten narrow raised cords; smaller secondary cords present between large cords; shoulder often encircled with variable number of small threadlike cords; when preserved, aragonitic layer inside aperture with eight to ten raised cords."

Type Specimen: HOLOTYPE: Length 38 mm, width 25 mm, Church Hill, Queen Anne's County, Maryland; originally in the collection of the Maryland Geological Survey, Baltimore, Maryland but now deposited in the collections of the Department of Paleobiology, National Museum of Natural History, Smithsonian Institution. The specimen shown here on Figure 3.9 A, B could be designated the lectotype if the holotype cannot be found in the Smithsonian collections.

Type Locality: Collected in a low cliff along South East Creek, near Church Hill, Queen Anne's County, Maryland.

Stratigraphic Range: Confined to the Plum Point Member (Shattuck Zone 10) of the Calvert Formation, Langhian Age of the early middle Miocene. The species is known from only three localities; Shattuck Zone 10 at Plum Point, Calvert County, Maryland (specimen illustrated here) and from the Plum Point Member of the Calvert Formation at Church Hill, Queen Anne's County, Maryland (the holotype); and the Pollack Farm site, Kent County, Delaware (Ward, 1998).

Etymology: Named for Church Hill, Maryland (from the Latin "ecclesiasticus", referring to a church).

Discussion: Martin (1904: plate LII, figure 9) illustrates a specimen of this distinctive species from "Church Hill, Maryland", but incorrectly identifies it as the older, Aquitanian Miocene *Ecphorosycon tampaensis*.

Chesathais donaldasheri Petuch, 1988
(Figure 3.10 A, B, C)

Description: (Reprinted from the original description; as *Chesathais lindae donaldasheri*; Petuch, 1988: 134-135, plate A3, figure D) "Shell distinctly vase-shaped, being wide across the shoulder and tapering abruptly toward anterior end; body whorl ornamented with 4 ribs, one bladelike in form and projecting upward along shoulder and anterior 3 being cordlike in form and concentrated around midbody; fifth obsolete cord present around juncture of body whorl and siphonal canal; intercostal areas heavily sculptured with numerous incised spiral threads; siphonal canal sculptured with numerous impressed threads; surface of upward-directed shoulder rib sculptured with 5 to 7 large spiral cords and numerous fine spiral threads; upward-pointing, bladelike shoulder rib produces canaliculated spire whorls; siphonal canal well developed, narrow; umbilicus poorly developed, narrow."
Type Specimens: HOLOTYPE: Length 37 mm (incomplete), Sandgates, St. Mary's County, Maryland, St. Leonards Member (Shattuck Zone 17), Choptank Formation, UF 21448; PARATYPE: length 22 mm, same locality and formation as the holotype, in the Petuch research collection.
Type Locality: From the basal bed of the low cliff along the Patuxent River at Sandgates, St. Mary's County, Maryland.
Stratigraphic Range: Confined to the St. Leonards Member (Shattuck Zone 17) of the Choptank Formation, early Serravallian Age of the middle Miocene. Known only from the type locality at Sandgates, St. Mary's County, Maryland. Specimens that weather out of the cliff accumulate at the cliff base and are often exposed on mud flats at low tide.
Etymology: Named for Donald F. Asher of Mechanicsville, St. Mary's County, Maryland, who collected the type material at Sandgates.
Discussion: *Chesathais donaldasheri* of the St. Leonards Member (Shattuck Zone 17) is similar to *C. lindae* from the suprajacent Drumcliff Member (Shattuck Zone 18) in being a broader, more vase-shaped shell with a lower spire and narrower siphonal canal. The mid-body ribs of *C. donaldasheri* are thinner and more cordlike than those of *C. lindae*, and the upward-directed shoulder rib is wider and more heavily-sculptured than that of the stratigraphically-higher descendant species. As in all the ecphoras collected in Zone 17 at Sandgates, all the known specimens of *C. donaldasheri* are poorly preserved and always collected in a fragmentary or fractured condition.

Chesathais lindae Petuch, 1988
(Figure 3.11 A, B)

Description: (Reprinted from the original description; Petuch, 1988: 9) "General shell form and structure as for genus; shell with 5 primary cords and 3 secondary cords; cord on shoulder raised, extended posteriorly, producing canaliculated spire whorls; fifth cord along base of siphonal canal smaller than other 4 cords; areas between cords squamose, slightly fimbriated on holotype."
Type Specimens: HOLOTYPE: length 31 mm, width 20 mm, north of Matoaka Cottages, St. Leonards, Calvert County, Maryland, USNM 424254; PARATYPES: length (broken) 17 mm, Drum Cliff, Patuxent River, St. Mary's County, Maryland, USNM 424255; 2 specimens, 59 mm and 36 mm, from the same locality as the holotype, in the Petuch research collection.
Type Locality: Along the Calvert Cliffs, 1 km north of Matoaka Cottages, St. Leonards, Calvert County, Maryland, in the Drumcliff Member (Shattuck Zone 18) of the Choptank Formation.
Stratigraphic Range: Confined to the Drumcliff Member (Shattuck Zone 18) of the Choptank Formation, early Serravallian Age of the middle Miocene.
Etymology: Named for Linda Joyce Petuch, wife of the senior author, who assisted him with field work when the holotype was collected.
Discussion: Martin (1904: plate LII, figure 10) illustrates a broken juvenile specimen of this rare shell from "Jones Wharf", but incorrectly identifies it as *Ecphorosycon tampaensis*, a much older species from the Aquitanian Miocene. This is the type species of the genus *Chesathais*.

Chesathais drumcliffensis Petuch, 1988
(Figure 3.12 A, B)

Description: (Reprinted from the original description; Petuch, 1988: 135-136 - originally described as a subspecies of *C. lindae*.) "Shell cylindrical in form, with rounded, sloping shoulder; body whorl ornamented with 4 ribs, one bladelike and directed upward along the shoulder, and anterior 3 being cordlike and concentrated around midbody; fifth small cord present around juncture of body whorl and siphonal canal; intercostal areas ornamented with one large spiral cord and numerous coarse spiral threads; siphonal canal sculptured with numerous raised cords; surface of upward-directed shoulder rib sculpture with 4 to 6 large spiral cords; upward-pointing, bladelike shoulder rib produces canaliculated spire whorls; siphonal canal well developed, narrow; umbilicus poorly-developed, narrow; interior of aperture with well-developed raised cords."
Type Specimens: HOLOTYPE: length 27 mm (incomplete), from the uppermost fossil shell bed (Shattuck Zone 19) at Drumcliff, Jones Wharf, St. Mary's County, Maryland, UF 21449; PARATYPE: length 32 mm, from the same locality and formation as the holotype, in the Petuch research collection.
Type Locality: Collected in the uppermost fossiliferous bed at Drumcliff (Jones Wharf), St. Mary's County, Maryland, along the Patuxent River.
Stratigraphic Range: Confined to the Boston Cliffs Member (Shattuck Zone 19) of the Choptank Formation, Serravallian Age of the middle Miocene.
Etymology: Named for Drumcliff, St. Mary's County, Maryland, along the Patuxent River, the type locality.
Discussion: This small, heavily-ornamented descendant species of *Chesathais lindae* represents the last member of the genus *Chesathais*. These rapaniform ecphoras became extinct at the end of Choptank time and are absent from all subsequent late Miocene and Pliocene formations.

Iconography of *Chesathais* Species

Figure 3.7 *Chesathais pungoensis* Petuch and Berschauer, new species
A, B= fragmentary holotype, length 21 mm. Collected in the PotashCorps Mine, Aurora, Beaufort County, North Carolina; from the Bonnerton Member, Pungo River Formation.

Figure 3.8 ***Chesathais whitfieldi*** **Petuch, 1988**
A, B= typical specimen, length 57 mm. Collected in Shattuck Zone 10 at Plum Point, Calvert County, Maryland; from the Plum Point Member of the Calvert Formation. Originally named from the Kirkwood Formation at Jericho, New Jersey.

Figure 3.9 ***Chesathais ecclesiasticus*** **(Dall, 1915)**
A, B= typical specimen, length 50 mm. Collected in Shattuck Zone 10 at Plum Point, Calvert County, Maryland; from the Plum Point Member, Calvert Formation. Originally named from Church Hill, Queen Anne County, eastern shore of Maryland (hence the name "ecclesiasticus").

Figure 3.10 ***Chesathais donaldasheri*** **Petuch, 1988**
A, B= fragmentary specimen, length 22 mm; **C=** fragmentary specimen, length 24 mm. Both specimens were collected in Shattuck Zone 17, in the basal bed at Sandgates, St. Mary's County, along the Patuxent River; from the St. Leonards Member, Choptank Formation.

Figure 3.11 ***Chesathais lindae*** **Petuch, 1988**
A, B= paratype specimen, length 36 mm. Collected in Shattuck Zone 18 at Matoaka Cottages, St. Leonards, Calvert County, Maryland; from the Drumcliff Member, Choptank Formation. This is the type species of the genus *Chesathais*.

Figure 3.12 ***Chesathais drumcliffensis*** **Petuch, 1988**
A, B= paratype specimen, length 32 mm. Collected in Shattuck Zone 19 (upper shell bed) at Drumcliff (Jones Wharf), along the Patuxent River, St. Mary's County, Maryland; from the Boston Cliffs Member, Choptank Formation, Serravallian Age of the middle Miocene. This is the last member of the genus *Chesathais* and the genus was extinct by the beginning of the St. Mary's Formation.

CHAPTER 4. The Genus *Trisecphora* Petuch, 1988 and Its Subgenera

With fifteen species and four separate subgenera, *Trisecphora* is the second-largest group of ecphoras and is the dominant genus during the Langhian Age of the Miocene (Calvert and Pungo River Formations). During that time, the stem stock wide-ribbed genus *Trisecphora (Trisecphora)* radiated into numerous habitats and evolved into three other distinct clades (considered subgenera in this work). These include *Trisecphora (Recurvecphora)*, *Trisecphora (Eccentrella)*, and *Trisecphora (Matoaka)* and these are described in the following sections. Shattuck Zone 10 of the Plum Point Member, Calvert Formation is the only known stratigraphic unit that contains representative species from all four of these subgeneric groups. The early Langhian Age, then, can be seen to have been the evolutionary acme for the genus. The last group of this complex, *Trisecphora (Matoaka)*, survived until the end of late Serravallian time, becoming extinct at the end of Choptank Formation deposition.

SYSTEMATICS

Class Gastropoda Cuvier, 1797
Subclass Orthogastropoda Ponder and Lindberg, 1995
Superorder Caenogastropoda Cox, 1960
Order Sorbeoconcha Ponder and Lindberg, 1997
Infraorder Neogastropoda Wenz, 1938
Superfamily Muricoidea Rafinesque, 1815
Family Muricidae Rafinesque, 1815
Subfamily Ocenebrinae Cossmann, 1903
Tribe Ecphorini Petuch and Berschauer, new tribe
Genus *Trisecphora* Petuch, 1988 (originally described as a subgenus of *Ecphora*)

Diagnosis: Shells similar to *Ecphora* species, being globose, inflated, with wide apertures and turriculate spires; body whorls ornamented with only 3 thin, raised ribs; ribs with or without flanges, some species with ribs that are "T"-shaped in cross section; faint, obsolete, fourth rib, often cordlike, encircles base of siphonal canal; species often sculptured with raised spiral threads between ribs and siphonal canals; last whorl disconnected from previous whorls in some species.
Type Species: *"Ecphora" tricostata* (Martin, 1904), Plum Point Member (Shattuck Zone 10) of the Calvert Formation of Maryland and the Pungo River Formation of North Carolina, Langhian Age of the early middle Miocene.
Species in *Trisecphora*: *Trisecphora precursor* Petuch and Berschauer, n. sp., Belhaven Member of the Pungo River Formation; *Trisecphora tricostata* (Martin, 1904), Plum Point Member, Calvert Formation and Bonnerton Member, Pungo River Formation; *Trisecphora governorensis* Petuch and Berschauer, n. sp., Plum Point Member (Shattuck Zone 12), Calvert Formation; *Trisecphora scientistensis* (Petuch, 1992), Calvert Beach Member (Shattuck Zone 14), Calvert Formation.

Trisecphora precursor Petuch and Berschauer, new species
(Figure 4.1 A, B)

Description: Shell small for genus, cylindrical in outline, with moderately-elevated spire whorls; siphonal canal proportionally long and narrow, encircled with 14 strong spiral cords; body whorl ornamented with 3 large, moderately wide, evenly-spaced thick ribs; poorly-developed smaller fourth rib present around anterior end of body whorl, adjacent to anteriormost larger rib; 3 main ribs slightly flanged along edges, slightly "T"- shaped in cross-section; areas of body whorl between main ribs heavily sculptured with 8-10 strong, closely-packed spiral cords; spire whorls encircled with 6-8 fainter spiral cords; aperture roughly circular in shape, with slightly raised parietal shield; umbilicus narrow, open.
Type Specimens: HOLOTYPE: length 29.6 mm, width 20.4 mm, PRI 104774; PARATYPE: fragmentary specimen, length 22 mm, from the same locality as the holotype, in the research collection of the senior author.
Type Locality: In the basal bed of the PotashCorps Phosphate Mine, Aurora, Beaufort County, North Carolina.
Stratigraphic Range: Restricted to the Belhaven Member of the Pungo River Formation, late Burdigalian Age of the Miocene.
Etymology: Named in reference to this new species being the first member of the true three-ribbed ecphoras (*Trisecphora* sensu stricto), presaging the larger and more advanced species from the Calvert and Choptank Formations of the Maryland Miocene.
Discussion: This new species is the oldest-known three-ribbed ecphora and represents the stem stock for the subsequent *Trisecphora* species radiations. It differs from its descendant species, *E. tricostata*, in being a smaller, much more cylindrical and less-inflated shell with much thinner ribs.

Trisecphora tricostata (Martin, 1904)
(Figure 4.2 A, B, C, D)

Description: (Reprinted from the original description, as *Ecphora tricostata*; Martin, 1904: 209-210) "Shell of moderate size, thin, chitinous, loosely coiled, six-whorled; nucleus smooth, elevated, conical three-whorled; spire sharp, closely wound on the lower and smaller of three prominent elevated revolving ribs; longitudinal wrinkles most prominent on the early whorls; body whorl large, with three very prominent elevated revolving ribs with a fourth rudimentary one below it on the largest specimens; spaces between the elevated ribs and below them with many impressed revolving lines between with the surface is gently convex; umbilicus widely flaring in the largest specimens, almost absent in the young."
Type Specimens: HOLOTYPE: length 40 mm, width 36 mm, from Shattuck Zone 10, Plum Point Member of the Calvert Formation at Plum Point, Calvert County, Maryland (holotype illustrated in Martin, 1904: Plate LII, Figure 5); originally in the collection of the Maryland Geological Survey, Baltimore, Maryland and now transferred to the Department of Paleobiology, National Museum of Natural History, Smithsonian Institution. The specimen illustrated here on Figure 4.2 A, B could be designated the lectotype if the holotype cannot be found in the Smithsonian collections.
Type Locality: In the basal bed exposed at Plum Point, Calvert County, Maryland.
Stratigraphic Range: Restricted to Shattuck Zones 10 and 11 of the Plum Point Member, Calvert Formation, Langhian Age, Middle Miocene.
Etymology: Named in reference to the three prominent spiral ribs that characterize this species.
Discussion: This is the type of the genus *Trisecphora s.s.* and the shell embodies all the important characteristics of the three-ribbed species group.

Trisecphora governorensis Petuch and Berschauer, new species
(Figure 4.3 A, B)

Description: Shell distinctly cylindrical in shape, only slightly inflated, proportionally narrow and laterally compressed; shoulder sharply angled; subsutural area flattened, planar; spire whorls stepped but proportionally low, rounded in profile; body whorl ornamented with 3 wide, flattened ribs that are broadly "T"-shaped in cross section; anterior two ribs sculptured with 1 or 2 faint, shallow impressed spiral threads; shoulder rib (posteriormost rib) wider than other two ribs, sculpted with 4 incised spiral threads; large shoulder rib projecting upward (posteriorward), bladelike in form and producing distinctly canaliculated spire whorls; small and poorly-developed cord present around base of body whorl; areas between ribs smooth, without spiral sculpture; siphonal canal proportionally long, narrow, ornamented with numerous very faint incised spiral threads; aperture oval; umbilicus narrow.
Type Specimen: HOLOTYPE: length 30.8 mm, width 22.9 mm, north of Governor Run, Calvert County, Maryland, Plum Point Member (Shattuck Zone 12) of the Calvert Formation, PRI 104775.
Type Locality: Collected in Shattuck Zone 12, 1.3 km north of Governor Run Pier, Calvert Cliffs, Calvert County, Maryland; from the upper bed of the Plum Point Member, Calvert Formation.
Stratigraphic Range: Confined to Shattuck Zone 12 of the Plum Point Member, Calvert Formation, late Langhian Age of the early Middle Miocene. Known only from the type locality.
Etymology: Named for Governor Run, the type locality.
Discussion: This Zone 12 descendant of the Zone 10 *Trisecphora tricostata* differs from its stratigraphically-lower ancestor in being a much more cylindrical and tightly-coiled shell with a narrower and more laterally-compressed shell outline, and in having proportionally-wider ribs whose edges are closer together, especially on the spire whorls. The wider shoulder rib pattern is retained in *T. scientistensis*, the descendant species from Shattuck Zone 14 (described next).

Trisecphora scientistensis (Petuch, 1992)
(Figure 4.4 A, B)

Description: (Reprinted from the original description, as *Ecphora (Trisecphora) scientistensis*; Petuch, 1992: 70-71, figure 10) "Shell cylindrical, slightly inflated in shape; shoulder sharply angled; subsutural area flattened, planar; spire whorls scalariform stepped; body whorl ornamented with 3 large cordlike ribs; ribs rounded on edges, sculptured with 1 or 2 faint, shallow impressed spiral threads; areas between ribs smooth, without spiral sculpture." Since the holotype was a damaged partial specimen, the siphonal canal and aperture structure was unknown at the time of the description. (Subsequent examination of new material, such as the specimen illustrated here, has shown that other important diagnostic characters are present on complete specimens. These are appended to the original description as follows): Shell with fourth subsidiary cord present on body whorl-siphonal canal juncture; subsidiary cord smaller than main cords, rounded, with no sculpture; 3 main ribs on body whorl unequal in size, with shoulder rib wider than other 2 ribs; shoulder rib flattened and "T"-shaped in cross section, ornamented with 4 deeply impressed spiral threads; siphonal canal proportionally long, broad, flexed at anterior end, ornamented with numerous incised spiral threads; umbilicus wide, open, and flaring; aperture proportionally large, oval in shape.
Type Specimen: HOLOTYPE: length 29 mm (incomplete and fragmentary), from Shattuck Zone 14, Calvert Formation at Scientists Cliffs, Calvert County, Maryland, UF 23799.
Type Locality: Collected in Shattuck Zone 14 at Scientists Cliffs, Calvert Cliffs, Calvert County, Maryland; in the Calvert Beach Member of the Calvert Formation.
Stratigraphic Range: Confined to Shattuck Zone 14 of the Calvert Beach Member, Calvert Formation, late Langhian Age of the early middle Miocene. Known only from Zone 14 beds at Scientists Cliffs and Governor Run, Calvert County, Maryland.
Etymology: Named for Scientists Cliffs, the type locality.
Discussion: *Trisecphora scientistensis* is the last of the wide-ribbed *Trisecphora* species found in the stratigraphic record; all subsequent three-ribbed taxa have thin, bladelike ribs and represent a separate evolutionary line (the subgenus *Matoaka*, described later in this chapter). Broken fragments of a thin-ribbed *Trisecphora* species, apparently an undescribed subspecies of the Shattuck Zone 10 *T. (Matoaka) prunicola*, have been collected along with *T. scientistensis* in Shattuck Zone 14, demonstrating that both thin-ribbed and wide-ribbed *Trisecphora* species occurred together during early Calvert Beach Member time.

Iconography of *Trisecphora* (sensu stricto) Species

Figure 4.1 ***Trisecphora precursor*** **Petuch and Berschauer, new species**
A, B= holotype, length 29.6 mm. Collected in the PotashCorps Phosphate Mine, Aurora, Beaufort County, North Carolina; from the Belhaven Member of the Pungo River Formation.

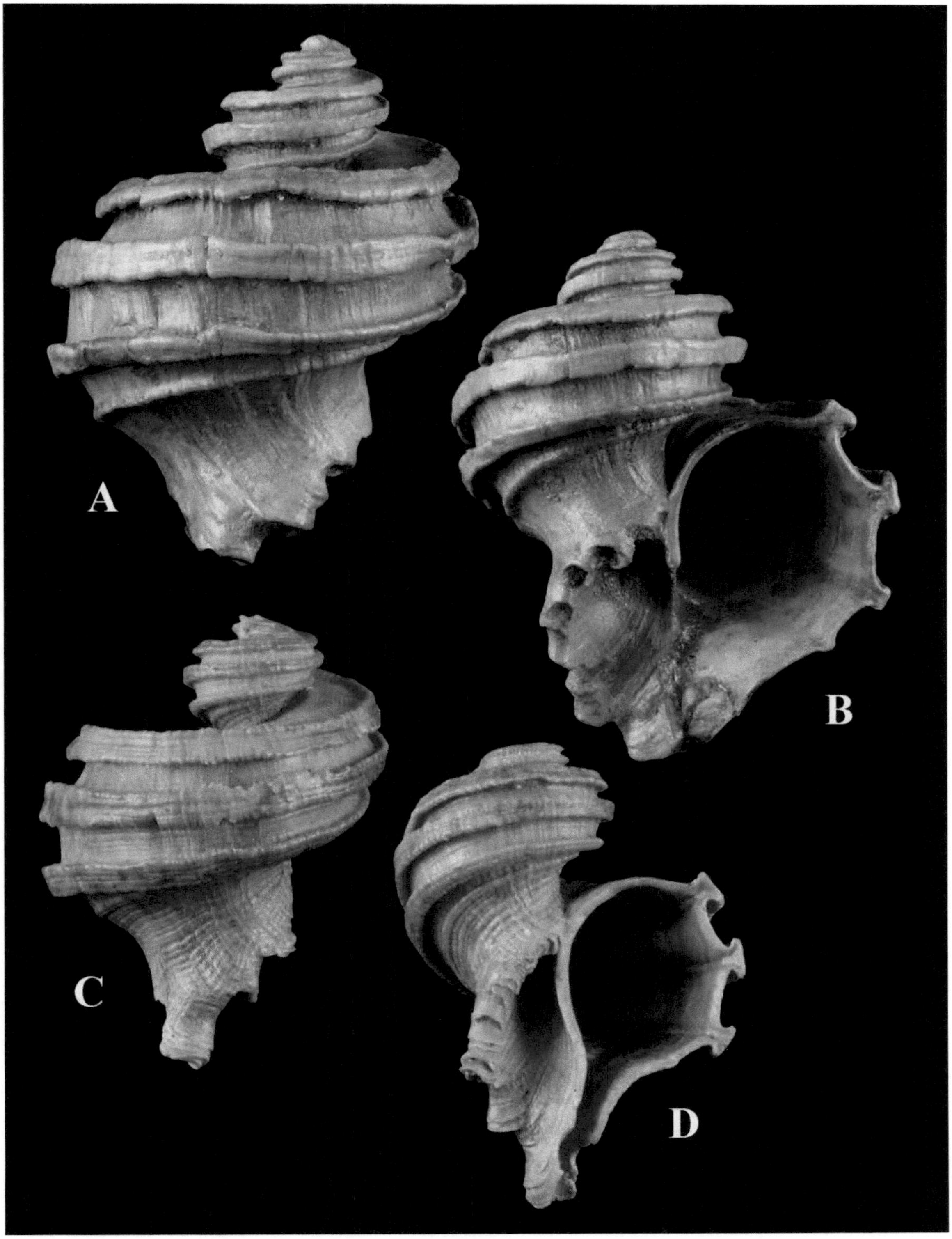

Figure 4.2 *Trisecphora tricostata* (Martin, 1904)
A, B= typical specimen, length 58 mm. Collected in Shattuck Zone 10 at Plum Point, Calvert County, Maryland; from the Plum Point Member of the Calvert Formation; **C, D**= typical specimen, length 35 mm. Collected in the PotashCorps Mine, Aurora, Beaufort County, North Carolina; from the upper beds of the Bonnerton Member, Pungo River Formation. This is the type species of the genus *Trisecphora s.s.*

Figure 4.3 *Trisecphora governorensis* Petuch and Berschauer, new species
A, B= holotype, length 32 mm. Collected in Shattuck Zone 12, 1.3 km north of Governor Run Pier, Calvert Cliffs, Calvert County, Maryland; from the upper bed of the Plum Point Member, Calvert Formation.

Figure 4.4 *Trisecphora scientistensis* (Petuch, 1992)
A, B= typical specimen, length 42 mm. Collected in Shattuck Zone 14, 1 km north of Governor Run, Calvert Cliffs, Calvert County, Maryland; from the lower bed of the Calvert Beach Member, Calvert Formation.

The Subgenus *Recurvecphora* Petuch and Berschauer, new subgenus

This small offshoot of the main *Trisecphora* evolutionary line, containing only three known species, was geologically short-lived, having been found only in the Langhian Miocene Pungo River and Calvert Formations. By the beginning of the Serravallian Age of the middle Miocene and the deposition of the Choptank Formation, the subgenus *Recurvecphora* became extinct. Unlike other groups of three-ribbed ecphoras, species in this subgenus characteristically have ribs that curve upward (posteriorward), producing a distinct cupped, or guttered, appearance. No other group within the Tribe Ecphorini has evolved ribs with this type of odd shape. In Shattuck Zone 10 of the Plum Point Member, Calvert Formation, the subgenus *Recurvecphora* occurs together with the *Trisecphora* subgenera *Eccentrella*, *Matoaka*, and *Trisecphora s.s.*, making this the largest and richest fauna of three-ribbed ecphoras found anywhere in the Chesapeake Miocene.

SYSTEMATICS
Class Gastropoda Cuvier, 1797
Subclass Orthogastropoda Ponder and Lindberg, 1995
Superorder Caenogastropoda Cox, 1960
Order Sorbeoconcha Ponder and Lindberg, 1997
Infraorder Neogastropoda Wenz, 1938
Superfamily Muricoidea Rafinesque, 1815
Family Muricidae Rafinesque, 1815
Subfamily Ocenebrinae Cossmann, 1903
Tribe Ecphorini Petuch and Berschauer, new tribe
Genus *Trisecphora* Petuch, 1988
Subgenus *Recurvecphora* Petuch and Berschauer, new subgenus

Diagnosis: Shells of average size for genus, distinctly vase-shaped and slightly inflated; spires stepped but proportionally low; body whorl ornamented with three large, prominent ribs, all of which curve upward (posteriorward), producing gutter-shaped, cuplike appearance; shoulder rib largest, projecting posteriorward to produce canaliculated spire whorls; ribs of body whorl ornamented with one or two deeply-incised spiral threads; siphonal canals proportionally long, often flexed or recurved; apertures large, oval; umbilici narrow.
Type Species: *Trisecphora (Recurvecphora) schmidti* (Petuch, 1988), Plum Point Member (Shattuck Zone 10), Calvert Formation of Maryland.
Species in *Recurvecphora*: *Trisecphora (Recurvecphora) recurvicostata* Petuch and Berschauer, n. sp., Belhaven Member, Pungo River Formation; *Trisecphora (Recurvecphora) schmidti* (Petuch, 1988), Plum Point Member (Shattuck Zone 10), Calvert Formation; *Trisecphora (Recurvecphora) martini* (Petuch, 1988), Calvert Beach Member (Shattuck Zone 14), Calvert Formation.

Trisecphora (Recurvecphora) recurvicostata Petuch and Berschauer, new species (Figure 4.5 A, B)

Description: Shell small for genus and subgenus, inflated, broadly vase-shaped, wide across shoulder; spire and spire whorls low, flattened; body whorl ornamented with 3 wide, flattened ribs, with 1-2 thin incised spiral threads on each rib; edges of ribs jagged and slightly squamose, directed upward (posteriorward), producing distinctive cupped appearance; small rounded cord present around base of body whorl; shoulder rib projecting above subsutural area, producing canaliculate spire whorls; areas between ribs and siphonal canal heavily ornamented with numerous fine, raised spiral threads; siphonal canal proportionally large and broad; umbilicus wide, well developed, slightly flaring.
Type Specimens: HOLOTYPE: length 18.4 mm, width 15.8 mm, Belhaven Member of the Pungo River Formation, at Aurora, North Carolina; PRI 104776.
Type Locality: In a block of phosphatic limestone from the lowest exposures in the Lee Creek Phosphate Mine (now PotashCorps Phosphate Mine), Aurora, Beaufort County, North Carolina; Belhaven Member, Pungo River Formation.
Stratigraphic Range: At present, known only from the Belhaven Member of the Pungo River Formation, early Langhian Age of the early middle Miocene. The species may eventually be recognized from the lowest beds of the Fairhaven and Plum Point Members, Calvert Formation (Shattuck Zones 2 and 4), but the poor preservation of specimens from those units makes accurate identification difficult, if not impossible.
Etymology: Named in reference to the distinctive, upward-curving structure of the ribs.
Discussion: Although superficially similar to the younger and stratigraphically-higher *Trisecphora (Recurvecphora) schmidti* (discussed next), this new Pungo River species differs primarily in having a more vase-shaped shell with a much wider shoulder, and in having the posterior two ribs being closer to each other than they are to the anterior rib, leaving a wide gap around the shell midbody.

Trisecphora (Recurvecphora) schmidti (Petuch, 1988)
(Figure 4.6 A, B)

Description: (Reprinted from the original description, as *Ecphora (Trisecphora) schmidti*; Petuch, 1988: 133-134, plate A3, figure A) "Shell inflated, distinctly vase-shaped; spire and spire whorls low, flattened; body whorl ornamented with 3 wide, flattened ribs; ribs directed upward (posteriorward), producing cupped appearance; shoulder rib projecting above subsutural area, producing canaliculate spire whorls; intercostal areas and siphonal canal heavily sculptured with numerous fine, raised spiral threads; umbilicus wide, well developed."
Type Specimens: HOLOTYPE: length 32 mm, in Shattuck Zone 10 of the Plum Point Member, Calvert Formation, at Plum Point, Calvert County, Maryland, UF 20522; PARATYPE: length 30 mm, from the same locality and formation as the holotype, in the Petuch research collection.
Type Locality: In the basal bed of a low cliff at Plum Point, Calvert County, Calvert Cliffs, Maryland; Shattuck Zone 10 of the Plum Point Member, Calvert Formation.
Stratigraphic Range: Restricted to Shattuck Zone 10 of the Plum Point Member, Calvert Formation, Langhian Age, Middle Miocene. Known only from the type locality at Plum Point.
Etymology: Named for Robert Schmidt, who collected the holotype at Plum Point.
Discussion: This distinctive large species, with its upturned ribs, is relatively common in the Zone 10 exposures at Plum Point, where it occurs together with three other *Trisecphora* species; *T. (Trisecphora) tricostata*, *Trisecphora (Eccentrella) eccentrica*, and *T. (Matoaka)*, making this the richest three-ribbed ecphora fauna found at any single locality.

Trisecphora (Recurvecphora) martini (Petuch, 1988)
(Figure 4.7 A, B)

Description: (Reprinted from the original description, as *Ecphora (Trisecphora) martini*; Petuch, 1988e: 26-28) "Shell vase-shaped, narrower at anterior end, flaring toward shoulder, body whorl ornamented with 3 main ribs and fourth auxiliary, weaker rib around base of shell; main ribs wide, bisected by impressed spiral groove; main ribs shaped like lopsided "T" in cross section, with top part of "T" extending and curving upward to produce cupped, channeled shelf on upper (posteriorward) surface of rib; shoulder rib curving upward, beyond plane of subsutural area, producing canaliculated spire whorls; auxiliary fourth rib bisected by groove on holotype but not on paratype, variable in size and development, but always without an upward curving flange and without cupped upper surface; main intersected by numerous vertical growth lines, giving ribs squamose appearance; areas between ribs smooth, without spiral sculpture; second rib (from shoulder, middle of 3 main ribs) extending outward from body farther than other ribs."

Type Specimens: HOLOTYPE: length 30 mm, width 29 mm, from approximately 100 m north of Governor Run Pier, Calvert Cliffs, Calvert County, Maryland, in the Calvert Beach Member (Shattuck Zone 14) of the Calvert Formation, USNM 427956; PARATYPE: length 25 mm, from the same locality and formation as the holotype, in the Petuch research collection.

Type Locality: Collected in Shattuck Zone 14 of the Calvert Beach Member, Calvert Formation, Calvert Cliffs, approximately 100 m north of Governor Run Pier, Calvert County, Maryland.

Stratigraphic Range: Confined to Shattuck Zone 14 of the Calvert Beach Member, Calvert Formation, late Langhian Age of the early middle Miocene. Known only from the area around Governor Run, Calvert County, Maryland.

Etymology: Named for George C. Martin, pioneer Maryland paleontologist, who was the first to complete a monographic work on the Miocene mollusks of Chesapeake Bay in 1904.

Discussion: Of the three species in the subgenus *Recurvecphora*, *E. (Recurvecphora) martini* has the most ornate and most sculpted ribs, with deeply-incised spiral grooves on each rib and with tiny crenulations along the rib edges.

Iconography of *Recurvecphora* Species

Figure 4.5 *Trisecphora (Recurvecphora) recurvicostata* Petuch and Berschauer, new species
A, B= holotype, length 21 mm. Collected in the PotashCorps Mine, Aurora, Beaufort County, North Carolina; from the Belhaven Member, Pungo River Formation.

Figure 4.6 ***Trisecphora (Recurvecphora) schmidti*** **(Petuch, 1988)**
A, B= paratype specimen, length 30 mm. Collected in Shattuck Zone 10 at Plum Point, Calvert County, Maryland; from the Plum Point Member, Calvert Formation. This is the type species of the subgenus *Recurvecphora*.

Figure 4.7 ***Trisecphora (Recurvecphora) martini*** **(Petuch, 1988)**
A, B= paratype specimen, length 25 mm. Collected in Shattuck Zone 14 at Governor Run Pier, Calvert Cliffs, Calvert County, Maryland; from the upper bed of the Plum Point Member, Calvert Formation.

The Subgenus *Eccentrella* Petuch and Berschauer, new subgenus

This small subgenus, containing only two species, represents a special offshoot of *Trisecphora* where the shell whorls have begun to uncoil and detach from each other. This morphology of detached whorls reaches its peak in *Trisecphora (Eccentrella) eccentrica*, where the spire whorls are highly extracted and projecting, forming a vermiform shell morphology like that seen in many turritellid gastropods of the genus *Vermicularia*. The subgenus was geologically short-lived, occurring only during early Langhian time and being found only in the Pungo River and Calvert Formations (only the Plum Point Member). The uncoiled shell structure may have been an adaptation for living on sea floors composed of fine mud and silt.

SYSTEMATICS
Class Gastropoda Cuvier, 1797
Subclass Orthogastropoda Ponder and Lindberg, 1995
Superorder Caenogastropoda Cox, 1960
Order Sorbeoconcha Ponder and Lindberg, 1997
Infraorder Neogastropoda Wenz, 1938
Superfamily Muricoidea Rafinesque, 1815
Family Muricidae Rafinesque, 1815
Subfamily Ocenebrinae Cossmann, 1903
Tribe Ecphorini Petuch and Berschauer, new tribe
Genus *Trisecphora* Petuch, 1988
Subgenus *Eccentrella* Petuch and Berschauer, new subgenus

Diagnosis: Shells of average size for genus, vase-shaped, with sharply angled shoulders; body whorl ornamented with three large, prominent rounded ribs which are slightly "T"-shaped in cross section; body and spire whorls characteristically become detached from each other, often uncoiling in vermiform fashion; spire whorls can be extremely protracted and exerted; siphonal canals proportionally large, broad; apertures large, oval in shape; umbilici large, open, flaring.
Type Species: *Trisecphora (Eccentrella) eccentrica* (Petuch, 1988), Plum Point Member (Shattuck Zone 10), Calvert Formation of Maryland.
Species in *Eccentrella*: *Trisecphora (Eccentrella) chamnessi* (Petuch, 1988), Belhaven Member, Pungo River Formation and Plum Point Member (Shattuck Zone 4), Calvert Formation; *Trisecphora (Eccentrella) eccentrica* (Petuch, 1988), Plum Point Member (Shattuck Zone 10), Calvert Formation.

Trisecphora (Eccentrella) chamnessi (Petuch, 1988)
(Figure 4.8 A, B, C)

Description: (Reprinted from the original description, as *Ecphora (Trisecphora) chamnessi*; Petuch, 1988e: 50-51, figure 19) "Shell compressed, discoidal in shape, with low spire; three main ribs of body whorl thick, rounded, cordlike; central rib projects farther from body whorl than other two ribs; small, almost obsolete, fourth rib present around base of siphonal canal; body and spire whorls heavily sculptured with fine, raised spiral threads between ribs and on siphonal canal; umbilicus narrow, flaring, well developed."

Type Specimens: HOLOTYPE: length 16 mm, diameter 19 mm, from the Belhaven Member of the Pungo River Formation exposed at the bottom of the Lee Creek Phosphate Mine (now PotashCorps Phosphate Mine), Aurora, Beaufort County, North Carolina, UF 20510; PARATYPE: length 49 mm, same locality and formation as the holotype, in the Petuch research collection.

Type Locality: In the basal beds exposed in the PotashCorps Phosphate Mine at Aurora, Beaufort County, North Carolina, Belhaven Member of the Pungo River Formation.

Stratigraphic Range: Confined to beds of late Burdigalian Age-early Langhian Age of the early middle Miocene and known only from the Belhaven Member of the Pungo River Formation of North Carolina and the basal bed of the Plum Point Member (Shattuck Zone 4) of the Calvert Formation of Maryland.

Etymology: Named for Ralph S. Chamness, Chief Geologist for the Texasgulf Chemicals Lee Creek Mine (now PotashCorps Phosphate Mine), Aurora, North Carolina, who kindly guided the senior author to rich exposures of the Pungo River Formation while studying the biostratigraphy of the mine.

Discussion: Although having distinctly uncoiled and unattached whorls like the younger and stratigraphically-higher *T. (Eccentrella) eccentrica*, the older Pungo River *T. (Eccentrella) chamnessi* has a much lower spire that is never as protracted and scalariform as its Shattuck Zone 10 descendant. Crushed and badly deformed specimens of *T. (Eccentrella) chamnessi* (one shown here on Figure 4.8C) are frequently encountered in Shattuck Zone 4 near Chesapeake Beach, Calvert County, demonstrating that the species was present in both Maryland and North Carolina during early Langhian time.

Trisecphora (Eccentrella) eccentrica (Petuch, 1988)
(Figures 4.9 A, B; 4.10)

Description: (Reprinted from the original description, as *Ecphora (Trisecphora) eccentrica*; Petuch 1988e: 132-133) “Shell inflated, distinctly vase-shaped, with sloping shoulder; body and spire whorls ornamented with 3 large, thick, low cords; cords slightly flattened along edges; intercostal areas sculptured with very fine, raised spiral threads, giving shell silky appearance; threads coarser on siphonal canal; shells consistently uncoiled, with spire whorls extremely protracted, scalariform, separated from each other; umbilicus wide, well developed; siphonal canal broad at body whorl juncture, tapering to narrow anterior tip.”
Type Specimens: HOLOTYPE: length 69 mm (missing spire whorls), from Shattuck Zone, Plum Point Member of the Calvert Formation, at Plum Point, Calvert County, northern Calvert Cliffs, Maryland, UF 20521; PARATYPE: length 56 mm, from the same locality and formation as the holotype, in the Petuch research collection.
Type Locality: From the lowest bed of the Calvert Cliffs at Plum Point, Calvert County, Maryland, Plum Point Member (Shattuck Zone 10) of the Calvert Formation.
Stratigraphic Range: Confined to Shattuck Zone 10 of the Plum Point Member, Calvert Formation, Langhian Age of the early middle Miocene. Known only from the Plum Point Member exposures along the Calvert Cliffs at, or near, Plum Point, Calvert County, Maryland.
Etymology: Named in reference to the eccentric, uncoiled appearance that characterizes this unusual ecphora.
Discussion: With its large shell and bizarrely uncoiled and projecting spire whorls, *Trisecphora (Eccentrella) eccentrica* is one of the most beautiful and unusual of all known ecphoras, especially those known from the Calvert Formation. Due to its structurally-fragile shell and uncoiled whorls, *T. (Eccentrella) eccentrica* is almost always collected in a broken state, often only as fragments (like the holotype). The best way to preserve this fragile shell, with its disconnected spire whorls, is to leave the entire specimen embedded in the matrix within which it was found, like the specimen shown here on Figure 4.10.

Iconography of *Eccentrella* Species

Figure 4.8 *Trisecphora (Eccentrella) chamnessi* (Petuch, 1988)
A, B= paratype specimen, length 49 mm. Collected in the PotashCorps Mine, Aurora, Beaufort County, North Carolina; from the Belhaven Member, Pungo River Formation. **C=** crushed and deformed specimen, length 53 mm. Collected in Shattuck Zone 4 (the gryphaeid oyster *Hyotissa percrassa* bed) at the base of the cliff at Chesapeake Beach, Calvert County, Maryland; from the lowest beds of the Plum Point Member, Calvert Formation.

Figure 4.9 *Trisecphora (Eccentrella) eccentrica* (Petuch, 1988)
A, B= paratype specimen showing the characteristic uncoiled shell and disconnected spire whorls, length 56 mm. Collected in Shattuck Zone 10 at Plum Point, Calvert County, Maryland; from the Plum Point Member, Calvert Formation. This is the type of the subgenus *Eccentrella*.

Figure 4.10 *Trisecphora (Eccentrella) eccentrica* (Petuch, 1988)
Typical specimen, length 74 mm, embedded in a block of sand matrix containing characteristic Plum Point Member mollusks such as the naticid *Euspirella tuomeyi* (Whitfield, 1894) (left side of the body whorl), the cochliopid *Carinorbis dalli* (Whitfield, 1894) (right side of the siphonal canal), fragments of the turrid *Calverturris bellacrenata* (Conrad, 1841) and the turritellid *Mariacolpus prunicola* Petuch and Drolshagen, 2010, and numerous large fragments of the glycymerid bivalve *Glycymeris parilis* (Conrad, 1843). Collected from an exposed wall of the Plum Point Member (Shattuck Zone 10), Calvert Formation, at Plum Point, Calvert County, Maryland.

The Subgenus *Matoaka* Petuch and Berschauer, new subgenus

This subgenus of three-ribbed ecphoras, with its thin bladelike ribs, constitutes the largest species group found within the genus *Trisecphora*. Members of this large subgenus were also the last species of three-ribbed ecphoras to have lived, becoming extinct at the end of Choptank Formation deposition (Serravallian Age, Miocene).

SYSTEMATICS
Class Gastropoda Cuvier, 1797
Subclass Orthogastropoda Ponder and Lindberg, 1995
Superorder Caenogastropoda Cox, 1960
Order Sorbeoconcha Ponder and Lindberg, 1997
Infraorder Neogastropoda Wenz, 1938
Superfamily Muricoidea Rafinesque, 1815
Family Muricidae Rafinesque, 1815
Subfamily Ocenebrinae Cossmann, 1903
Tribe Ecphorini Petuch and Berschauer, new tribe
Genus *Trisecphora* Petuch, 1988
Subgenus *Matoaka* Petuch and Berschauer, new subgenus

Diagnosis: Shells distinctly vase-shaped, with rounded body whorls and elongated siphonal canals; body whorls characteristically ornamented with three very thin, bladelike ribs, with some being low and rounded and others projecting from the body whorl and undulating; fourth small, cordlike, rib present on some species, at the juncture of the body whorl and siphonal canal; apertures oval; umbilici narrow.
Type Species: *Trisecphora (Matoaka) patuxentia* (Petuch, 1988), Drumcliff Member (Shattuck Zone 18), Choptank Formation of Maryland.
Species in *Matoaka*: *Trisecphora (Matoaka) prunicola* (Petuch, 1988), Shattuck Zone 10, Plum Point Member, Calvert Formation; *Trisecphora (Matoaka) carolinensis* (Petuch, 1988), Bonnerton Member, Pungo River Formation; *Trisecphora (Matoaka) bartoni* Petuch and Drolshagen, 2010, Shattuck Zone 16, Calvert Beach Member, Calvert Formation; *Trisecphora (Matoaka) smithae* (Petuch, 1988), Shattuck Zone 17, St. Leonard Member, Choptank Formation; *Trisecphora (Matoaka) patuxentia* (Petuch, 1988), Shattuck Zone 18, Drumcliff Member, Choptank Formation; *Trisecphora (Matoaka) shattucki* (Petuch, 1988), Shattuck Zone 19, Boston Cliffs Member, Choptank Formation.
Discussion: The new subgenus name honors Matoaka, daughter of the regional Powhatan of the tribal confederation (Tsenacommacah) of coastal Virginia and Chesapeake Bay, who is better known in modern times as Pocahontas.

***Trisecphora (Matoaka) prunicola* (Petuch, 1988)**
(Figure 4.11 A, B)

Description: (Reprinted from the original description, as *Ecphora (Trisecphora) prunicola*; Petuch, 1988e: 15) "Shell small for genus and subgenus, thin, fragile, widely inflated; shoulder strongly angled, spire protracted; 3 ribs thin, bladelike, projecting from body whorl; last whorl on mature specimens disconnected, uncoiled; siphonal canal well-developed, sculptured with fine spiral threads; umbilicus well-developed but narrow."
Type Specimens: HOLOTYPE: length 26 mm, width 24 mm, Shattuck Zone 10 at Plum Point, Calvert Cliffs, Calvert County, Maryland, Plum Point Member of the Calvert Formation, USNM 424259; PARATYPE: length 33 mm, same locality and stratigraphic unit as the holotype, in the research collection of the senior author.
Type Locality: In Shattuck Zone 10 at Plum Point, Calvert Cliffs, Calvert County, Maryland; in the base of a low cliff that borders the beach.
Stratigraphic Range: Confined to Shattuck Zone 10 of the Plum Point Member, Calvert Formation, Langhian Age, early middle Miocene.
Etymology: Named for Plum Point, Calvert County, Maryland.
Discussion: *Trisecphora (Matoaka) prunicola* is most similar to *T. (Matoaka) carolinensis* from the contemporaneous Pungo River Formation of North Carolina (described next) but differs in being a much more elongated and slender shell with a much higher, more protracted spire. On the Pungo River species, the central rib is much closer to the shoulder rib, producing a bi-costate appearance. On *T. (Matoaka) prunicola*, the three ribs are all equally separated from each other.

***Trisecphora (Matoaka) carolinensis* (Petuch, 1988)**
(Figure 4.12 A, B, C, D)

Description: (Reprinted from the original description, as *Ecphora (Trisecphora) prunicola carolinensis*; Petuch, 1988e: 133, plate A3, figure B) "Shell flattened, discoidal, with rounded shoulder; whorl ornamented with 3 very thin, bladelike ribs; shoulder and midbody ribs closer together than midbody and anterior ribs; intercostal areas and siphonal canal heavily sculptured with coarse spiral threads and thin cords; umbilicus well developed."

Type Specimens: HOLOTYPE: length 16 mm (incomplete), from the Bonnerton Member of the Pungo River Formation, early Langhian Age of the Miocene; PARATYPE: length 30 mm, same locality and formation as the holotype, in the Petuch research collection.

Type Locality: In the lowest beds in the Lee Creek Phosphate Mine (now PotashCorp Phosphate Mine), Aurora, Beaufort County, North Carolina; Bonnerton Member of the Pungo River Formation.

Stratigraphic Range: Confined to the Bonnerton Member of the Pungo River Formation, early Langhian Age of the early middle Miocene. Known only from the Pungo River exposures at Aurora, North Carolina.

Etymology: Named for the State of North Carolina, southeastern United States.

Discussion: The shoulder rib of *Trisecphora (Matoaka) carolinensis* is much smaller than the other two ribs on the body whorl, and is closer to the mid-body rib than the mid-body rib is to the anterior rib. The asymmetry of the ribs gives this ecphora a distinct bicostate appearance, looking like a two-ribbed shell when seen from the apertural (ventral) side.

Trisecphora (Matoaka) bartoni Petuch and Drolshagen, 2010 (Figure 4.13 A, B, C, D)

Description: (Reprinted from the original description; Petuch and Drolshagen, 2010: 127, figure 4.6 G, H) "Shell of average size for genus, distinctly vase-shaped, with wide shoulder and midbody area tapering gradually to elongated siphonal canal; spire whorls scalariform, with flat subsutural areas; ribs three in number, thin, bladelike, with shoulder rib projecting upward, producing canaliculated subsutural area; fourth small, undeveloped rib present around base of siphonal canal on some specimens; aperture proportionally large, oval in shape; umbilicus narrow and deep; siphonal canal ornamented with numerous faint spiral threads."
Type Specimens: HOLOTYPE: length (incomplete) 34 mm, width 27 mm, from Shattuck Zone 16 of the Calvert Formation, in cliffs 0.5 km north of Governor Run, Calvert County, Calvert Cliffs, Maryland. PRI 9920; PARATYPES: 2 specimens, length 23 mm and 30 mm, from the same locality and formation as the holotype, in the Petuch research collection.
Type Locality: Found in Shattuck Zone 16 in the Calvert Cliffs, 0.5 km north of Governor Run, Calvert County, Maryland; Calvert Beach Member of the Calvert Formation.
Stratigraphic Range: Restricted to Shattuck Zone 16 of the Calvert Beach Member, Calvert Formation, late Langhian Age of the early middle Miocene.
Etymology: Named for Matthew Barton of Boca Raton, Florida.
Discussion: With its distinctive tapering, vase-shaped shell, *Trisecphora (Matoaka) bartoni* cannot be confused with any of the other three-ribbed ecphora species.

Trisecphora (Matoaka) smithae (Petuch, 1988)
(Figure 4.14 A, B, C)

Description: (Reprinted from the original description, as *Ecphora (Trisecphora) smithae*; Petuch, 1988e: 28) "Shell greatly inflated, tumid, globose, with rounded shoulder; body whorl with 3 thick, low, cordlike ribs; ribs unornamented, evenly spaced, rounded in cross section; areas between ribs smooth and unornamented; few faint raised threads sometimes present on siphonal canal; umbilicus large, open, widely-flaring; whorls attached, tightly coiled."

Type Specimens: HOLOTYPE: length 42 mm, width 41 mm, in Shattuck Zone 17 at Matoaka Cottages, St. Leonards, Calvert County, Maryland, in the St. Leonard Member of the Choptank Formation, USNM 427958; PARATYPES: length 40 mm (fragmentary), same locality and formation as the holotype, USNM 427959; length 78 mm, same locality and formation as the holotype, in the Petuch research collection; length 44 mm, from Shattuck Zone 17 at Sandgates, St. Mary's County, Maryland, along the Patuxent River, in the Petuch research collection.

Type Locality: From Shattuck Zone 17 in the Calvert Cliffs at Matoaka Cottages, St. Leonards, Calvert County, Maryland.

Stratigraphic Range: Confined to the early Serravallian Age of the middle Miocene, in the St. Leonard Member (Shattuck Zone 17) of the Choptank Formation. Specimens are abundant at several Zone 17 sites along the Calvert Cliffs near St. Leonards, Maryland, and also along the Patuxent River at Sandgates, St. Mary's County, Maryland.

Etymology: Named for Connie Smith, owner of Matoaka Cottages, St. Leonard, Calvert County, Maryland, who generously allowed paleontologists access to several important Choptank research sites on her property.

Discussion: *Trisecphora (Matoaka) smithae* is the largest and most inflated species of its subgenus, with some specimens reaching the size of a small grapefruit.

Trisecphora (Matoaka) patuxentia (Petuch, 1988)
(Figure 4.15 A, B, C, D)

Description: (Reprinted from the original description, as *Ecphora (Trisecphora) patuxentia*; Petuch, 1988e: 8) "General form as for genus; 3 ribs thin, bladelike, rounded on edges; ribs not straight but undulating; areas between ribs sculpted with raised spiral threads; central rib more elevated than anterior and posterior ribs."
Type Specimens: HOLOTYPE: length 53 mm, width 39 mm, USNM 424252; PARATYPES: length 34 mm, same locality as the holotype, USNM 424253; 2 specimens, lengths 77 mm and 37 mm, same locality as the holotype, in the research collection of the senior author.
Type Locality: Along the northern side of Patuxent River, at the base of a cliff 10 km north of the Patuxent River Bridge, St. Mary's County, Maryland.
Stratigraphic Range: Confined to the Drumcliff Member (Shattuck Zone 18) of the Choptank Formation, Serravallian Age, Middle Miocene.
Etymology: Named for the Patuxent River of Maryland.
Discussion: Although the original description states that there are "raised spiral threads" between the ribs, these are not prominent and are nearly obsolete. On well-preserved specimens with little erosion, these threads are extremely fine and are not nearly as coarse or prominent as those seen on the older *Trisecphora (Matoaka) prunicola* from the Plum Point Member of the Calvert Formation. This Serravallian-aged species is most similar to the older *T. (Matoaka) smithae* from Shattuck Zone 17 of the St. Leonard Member, but differs in having the shoulder rib project upward (posteriorward). The ribs of *T. (Matoaka) patuxentia* characteristically are not planar, but undulate to some degree, with some specimens having extreme ripples that extend beyond the plane of the rib (such as in the specimen shown here on Figure 4.15D). This is the type species of the subgenus *Matoaka*.

Trisecphora (Matoaka) shattucki (Petuch, 1988)
(Figure 4.16 A, B, C)

Description: (Reprinted from the original description, as *Ecphora (Trisecphora) patuxentia shattucki*; Petuch, 1988e: 53-54: figure 25) "Shell cylindrical, elongated, with elevated spire; body whorl ornamented with three thick, cordlike ribs; shoulder rib on spire whorls projecting straight upward; entire spire, body whorl, and siphonal canal sculptured with fine raised cords and spiral threads; thin, low fourth rib present around base of siphonal canal; siphonal canal long, well-developed; umbilicus narrow."

Type Specimen: HOLOTYPE: length 35 mm, from the upper shell bed at Drumcliff, Jones Wharf, St. Mary's County, Maryland, UF 20511; PARATYPES: two specimens, lengths 25 mm and 25.3 mm, from the same locality as the holotype, in the Petuch research collection.

Type Locality: In the upper shell bed (Shattuck Zone 19) at Drumcliff, Jones Wharf, St. Mary's County, on the Patuxent River; Boston Cliffs Member of the Choptank Formation.

Stratigraphic Range: Confined to Shattuck Zone 19, Boston Cliffs Member of the Choptank Formation, late Serravallian Age of the middle Miocene.

Etymology: Named for George B. Shattuck, pioneer Maryland geologist and stratigrapher.

Discussion: The Shattuck Zone 19 *Trisecphora (Matoaka) shattucki* is morphologically closest to the older and stratigraphically-lower *T. (Matoaka) patuxentia* from Shattuck Zone 18, but differs in being a smaller, more cylindrical, and less inflated shell and in having much thicker, less bladelike ribs on the body whorl. This Zone 19 species is rare at the type locality and is the last species of three-ribbed ecphora found in the fossil record. Its rarity reflects the fact that the genus *Trisecphora* was dying out at the end of Serravallian time. No more three-ribbed ecphoras are present in the subsequent late Miocene and Pliocene formations.

Iconography of *Matoaka* Species

Figure 4.11 *Trisecphora (Matoaka) prunicola* (Petuch, 1988)
A, B= paratype specimen, length 33 mm, collected in Shattuck Zone 10 at Plum Point, Calvert County; from the Plum Point Member, Calvert Formation.

Figure 4.12 *Trisecphora (Matoaka) carolinensis* (Petuch, 1988)
A, B= paratype specimen, length 30 mm; **C, D**= typical specimen, length 19 mm. Both specimens collected in the PotashCorps Mine, Aurora, Beaufort County, North Carolina; from the Bonnerton Member, Pungo River Formation.

Figure 4.13 *Trisecphora (Matoaka) bartoni* Petuch and Drolshagen, 2010
A, B= paratype specimen, length 30 mm. Collected in Shattuck Zone 16, 0.5 km north of Governor Run, Calvert Cliffs, Calvert County, Maryland; from the upper bed of the Calvert Beach Member, Calvert Formation.

Figure 4.14 *Trisecphora (Matoaka) smithae* (Petuch, 1988)
A, B= paratype specimen, length 78 mm, collected in Shattuck Zone 17 at Matoaka Cottages, St. Leonards, Calvert County, Maryland; **C**= paratype specimen, length 44 mm, collected in Shattuck Zone 17 at Sandgates, along the Patuxent River, St. Mary's County, Maryland. Both specimens are from the St. Leonard Member, Choptank Formation.

Figure 4.15 *Trisecphora (Matoaka) patuxentia* (Petuch, 1988)
A, B= paratype specimen, length 37 mm; **C=** typical specimen, length 62 mm; **D=** specimen with well-developed undulating ribs, length 25 mm. All three specimens were collected in Shattuck Zone 18 at Sandgates, in a cliff along the Patuxent River, St. Mary's County, Maryland; from the Drumcliff Member of the Choptank Formation. This is the type species of the subgenus *Matoaka*.

Figure 4.16 *Trisecphora (Matoaka) shattucki* (Petuch, 1988)
A, B= paratype specimen, length 25 mm; **C=** paratype specimen, length 25.3 mm. Both specimens collected in Shattuck Zone 19 at Drumcliff (Jones Wharf), along the Patuxent River, St. Mary's County, Maryland. This was the last species of three-ribbed ecphora shell; the genus *Trisecphora* disappears from the fossil record after Shattuck Zone 19 and is not found in the St. Mary's or any other subsequent formation.

Ecphora roxaneae dalli Petuch and Berschauer, new subspecies, length 111 mm, from the Goose Creek Formation near Santee, South Carolina.

CHAPTER 5. The Genus *Ecphora* Conrad, 1843 and Its Subgenera

This genus represents the classic type of ecphora shell, as described by the pioneer American malacologists, Thomas Say and Timothy Conrad. With 13 known species and subspecies, *Ecphora sensu stricto* is the largest genus of ecphoras and is also the most wide-ranging geographically, being found in fossil beds from Maryland, Virginia, North and South Carolina, northwestern Florida, to extreme southern Florida (Miami). An augmented and expanded diagnosis of the genus is given here:

The Genus *Ecphora* (sensu stricto) Conrad, 1843

SYSTEMATICS
Class Gastropoda Cuvier, 1797
Subclass Orthogastropoda Ponder and Lindberg, 1995
Superorder Caenogastropoda Cox, 1960
Order Sorbeoconcha Ponder and Lindberg, 1997
Infraorder Neogastropoda Wenz, 1938
Superfamily Muricoidea Rafinesque, 1815
Family Muricidae Rafinesque, 1815
Subfamily Ocenebrinae Cossmann, 1903
Tribe Ecphorini Petuch, 1988
Genus *Ecphora* Conrad, 1843

Diagnosis: (Original description taken from Conrad (1843: 324): "Shell fusiform, ventricose, with revolving costae; aperture oval, suddenly contracting towards the base of the beak, which is reflected; labrum united with the labrum above; labrum reflected; margins of the beak so elevated as to form a profound very oblique channel towards the base of the shell." (Expanded description from Petuch, 1988:63): "Shell inflated, globose, subfusiform in shape: apertures wide, oval in shape; siphonal canals long, well developed; umbilici variable in size, generally well developed, but with some species having wide, flaring umbilici, others with narrow, slender umbilici; shells ornamented with four or five ribs on body whorl and two or three ribs on spire whorls; ribs variable in form, species-specific, with some being low, rounded, and cordlike while others have wide ribs that are "T"-shaped in cross section; areas between ribs having variable amounts of sculpture, with some species having smooth shells and others being sculptured with fine, raised spiral threads."
Type Species: *Ecphora quadricostata* (Say, 1824), Rushmere Member of the Yorktown Formation (originally "*Fusus 4-costatus*").
Species in *Ecphora*: *Ecphora wardi* Petuch, 1988, Bonnerton Member, Pungo River Formation and Plum Point Member (Shattuck Zone 10), Calvert Formation; *Ecphora mattinglyi* Petuch, 2004, Calvert Beach Member (Shattuck Zone 14), Calvert Formation; *Ecphora sandgatesensis* Petuch, 1988, St. Leonard Member (Shattuck Zone 17), Choptank Formation; *Ecphora meganae* Ward and Gilinsky, 1988, Drumcliff Member

(Shattuck Zone 18), Choptank Formation; *Ecphora amyae* Petuch and Drolshagen, 2010, Boston Cliffs Member (Shattuck Zone 19), Choptank Formation; *Ecphora kochi* Ward and Gilinsky, 1988, Cobham Bay Member, Eastover Formation and lowest bed of the Goose Creek Formation; *Ecphora roxaneae* Petuch, 1991, Sarasota Member, Murdock Formation; *Ecphora roxaneae dalli* Petuch and Berschauer, new subspecies, Goose Creek Formation; *Ecphora leecreekensis* Petuch, 1988, Sunken Meadow Member, Yorktown Formation; *Ecphora quadricostata* (Say, 1824), Rushmere Member, Yorktown Formation, Goose Creek Formation, and Jupiter Waterway Member, Murdock Formation; *Ecphora quadricostata striatula* Petuch, 1986, Buckingham Member, Tamiami Formation and Jackson Bluff Formation ("*Ecphora* Zone" lower bed); *Ecphora quadricostata umbilicata* (Wagner, 1839), Morgarts Beach Member, Yorktown Formation and Pinecrest Member (lowermost beds), Tamiami Formation; *Ecphora quadricostata rachelae* Petuch, 1988, Moore House Member, Yorktown Formation and Pinecrest Member (uppermost beds), Tamiami Formation.

Discussion: The genus *Ecphora s.s.* is a very conservative group, with little morphological change having occurred over the timespan of its existence. From the early Langhian Age of the Miocene to the mid-Piacenzian Age of the Pliocene, the only major difference between the species and subspecies is size, the amount of secondary ornamentation on the body whorl, and the strength of the grooves on the ribs.

Ecphora unnamed species, length 30 mm, from Shattuck Zone 11, Plum Point Member, Calvert Formation, south of Chesapeake Beach, Maryland. Known only from fragmentary specimens, this rare ecphora represents the evolutionary node where the subgenus *Powhatan* diverged from the main *Ecphora s.s.* lineage.

Ecphora wardi Petuch, 1988
(Figure 5.1 A, B, C, D)

Description: (Reprinted from the original description, as *Ecphora (Ecphora) wardi*, Petuch, 1988: 80-81, figure 35) "Shell large for genus, cylindrical, very elongated with high scalariform spire; body whorl ornamented with four thin, bladelike ribs; edge of ribs rounded, with single faint, impressed spiral thread; body whorl and spire whorls between ribs ornamented with numerous fine spiral threads and faint cords; siphonal canal very elongated, well developed, ornamented with spiral threads; umbilicus large, flaring; shoulder rib projecting upward, producing canaliculate spire whorls; body whorl of older adult specimens detached from rest of shell.

Type Specimens: HOLOTYPE: length 70.3 mm, from Shattuck Zone 10 at Camp Roosevelt, Calvert County, Maryland (illustrated in Ward and Gilinsky, 1988: 17, plate 1, figures 4 and 5, as "*Ecphora tricostata*"), USNM 405322; PARATYPE: length 55.6 mm, from Shattuck Zone 10 at Plum Point, Calvert County, Maryland, in the Petuch research collection.

Type Locality: Camp Roosevelt, Calvert County, Maryland, in Shattuck Zone 10 of the Plum Point Member, Calvert Formation.

Stratigraphic Range: Confined to the early Langhian Age of the early middle Miocene of Maryland and North Carolina, being found in the Plum Point Member (Shattuck Zone 10), Calvert Formation of Maryland and also the Bonnerton Member, Pungo River Formation of North Carolina.

Etymology: Named for Dr. Lauck W. ("Buck") Ward, formerly of the US Geological Survey and Virginia Museum of Natural History.

Discussion: This widespread Langhian ecphora is the first species to have four distinct ribs and is ancestral to all subsequent four-ribbed ecphoras. With its upwardly-projecting shoulder rib, *Ecphora wardi* is clearly ancestral to species such as *E. mattinglyi*, *E. meganae*, and *E. amyae* and is the first species of the nominate subgenus *Ecphora (Ecphora)*. Of the descendant species, *E. wardi* is most similar to *E. mattinglyi* from Shattuck Zone 14 of the Calvert Beach Member, Calvert County (described next), especially in general shell shape and spire height, but differs in having thinner and more bladelike ribs that are not as rounded and cordlike as those seen on the younger and stratigraphically-higher species. This superficial similarity led the senior author to include a specimen of the Zone 14 shell as a paratype in the original description of *E. wardi* (Petuch, 1988: figure 35 A, B). This specimen, illustrated next, was removed from the type lot of *E. wardi* and later became part of the type lot of a new descendant species, *E. mattinglyi* Petuch, 2004. Specimens of *E. wardi* from the Pungo River Formation of North Carolina are smaller than those from Maryland and may represent an undescribed subspecies.

Ecphora mattinglyi Petuch, 2004
(Figure 5.2 A, B)

Description: (Reprinted from the original description, Petuch, 2004:282, plate 24F, G) "Shell of average size for genus, distinctly cylindrical, with high protracted spire and proportionally-long siphonal canal; shoulder angled, subsutural area flattened; body whorl ornamented with 4 large, highly elevated spiral ribs; ribs flattened on edges, slightly "T"-shaped in cross section; rib on shoulder raised, bladelike, producing slightly canaliculated spire whorls; spire whorls with 3 large spiral ribs; siphonal canal ornamented with 10 low, faint, closely-packed spiral threads; umbilicus proportionally large, open, flaring."
Type Specimens: HOLOTYPE: length 56 mm, width 46 mm, Shattuck Zone 14, Calvert Beach Member of the Calvert Formation, near Chesapeake Beach, Calvert County, Maryland, AMNH FI-50675; PARATYPE: length 88 mm, from Shattuck Zone 14 of the Calvert Beach Member, Calvert Formation at Scientists Cliffs, Calvert County, Maryland.
Type Locality: Shattuck Zone 14 (Calvert Beach Member, Calvert Formation) at Scientists Cliffs, Calvert County, Maryland.
Stratigraphic Range: Confined to Shattuck Zone 14 of the Calvert Beach Formation, Calvert Formation, late Langhian Age, early middle Miocene.
Etymology: Named for Edward Mattingly of Owings, Maryland, who collected the holotype at Chesapeake Beach.
Discussion: *Ecphora mattinglyi* is most similar to the younger and stratigraphically-higher Zone 17 (St. Leonards Member) *E. sandgatesensis* (described next), but differs in being a larger shell with a more inflated body whorl, in having a proportionally higher and more exerted spire, and in having a wider and more rounded, thicker, less upturned, and less bladelike shoulder rib.

Ecphora sandgatesensis Petuch, 1988
(Figure 5.3 A, B, C, D)

Description: (Reprinted from the original description, as *Ecphora (Ecphora) meganae sandgatesensis*; Petuch, 1988: 130, plate A1, figures B, C) "Shell cylindrical in form, relatively straight-sided, shoulder wide; subsutural area flattened, broad; spire scalariform; body whorl ornamented with 4 cordlike ribs; shoulder rib more bladelike than body ribs, projecting upward to produce distinctly canaliculated spire whorls; three body ribs slightly "T"-shaped in cross section; areas between ribs ornamented with numerous thin, prominent, raised threads and cords; faint fifth rib sometimes present around base of siphonal canal; siphonal canal well developed, elongated, ornamented with fine raised threads and impressed sulci; umbilicus poorly developed, narrow."
Type Specimens: HOLOTYPE: length 29 mm (incomplete), in the basal bed of the cliff at Sandgates, St. Mary's County, Maryland, at the waterline of the Patuexent River, Shattuck Zone 17, St. Leonard Member, Choptank Formation, UF 20518; PARATYPES: 2 specimens, lengths 40 and 44 mm, from the same locality and formation as the holotype, in the Petuch research collection; 8 specimens, lengths 30-38 mm, from the same locality and formation as the holotype, in the Donald Asher collection.
Type Locality: Found in the basal bed of the cliff adjacent to the Sandgates dock, Sandgates, St. Mary's County, Maryland, at the waterline of the Patuxent River; from Shattuck Zone 17, St. Leonards Member, Choptank Formation.
Stratigraphic Range: Confined to the St. Leonards Member (Shattuck Zone 17) of the Choptank Formation, early Serravallian Age of the middle Miocene. Found in Shattuck Zone 17 beds along the Patuxent River and also along the Calvert Cliffs north of St. Leonards, Calvert County, Maryland.
Etymology: Named for Sandgates, St. Mary's County, Maryland, the type locality.
Discussion: The Shattuck Zone 17 *Ecphora sandgatesensis* differs from the younger and stratigraphically-higher descendant species, *E. meganae* from Zone 18, in being a smaller, more cylindrical shell with narrower and thinner ribs, and in having a better developed, wider upward-projecting shoulder rib.

Ecphora meganae Ward and Gilinsky, 1988
(Figure 5.4 A, B, C, D)

Description: (abbreviated description, taken from Ward and Gilinsky, 1988: 4-5) "Shell cylindrical in form with high, elevated spire; shoulder sharp-angled; ribs four in number, moderately wide, flattened along their edges; edges of ribs ornamented with one or two deeply impressed spiral threads; shoulder rib on spire whorls projects upward (posteriorward), producing canaliculated spire; areas between ribs heavily sculptured with fine spiral threads and cords; fifth, poorly developed rib often present around anterior of body whorl; umbilicus well developed, flaring."
Type Specimens: HOLOTYPE: length 64.6 mm, width 48.5 mm, in the lower bed at Drumcliff, Jones Wharf, St. Mary's County, Maryland, on the Patuxent River, Drumcliff Member (Shattuck Zone 18) of the Choptank Formation, USNM 405320; PARATYPES: (from the same locality and formation as the holotype) length 69.4 mm, USNM 424302; length 60.8 mm, USNM 424303.
Type Locality: The lower shell bed (Shattuck Zone 18) at Drumcliff, Jones Wharf, St. Mary's County, Maryland, along the Patuxent River.
Stratigraphic Range: Confined to the Drumcliff Member (Shattuck Zone 18) of the Choptank Formation, early Serravallian Age of the Miocene. The species is commonly encountered in all Zone 18 beds, especially at Matoaka Cottages and Governor Run, near St. Leonards, Calvert County, Maryland and at Drumcliff and Jones Wharf, St. Mary's County, along the Patuxent River.
Etymology: Named for Megan H. Jones, Geology Department, Old Dominion University.
Discussion: Ward and Gilinsky's original description (1988) of *Ecphora meganae* is very confusing, primarily because the authors included three separate species, in three different generic groups, within their species description. Besides the holotype of their *E. meganae*, which closely resembles the two specimens shown here on Figure 5.4, they also included specimens of *E. (Rhombecphora) rikeri* and *Trisecphora (Matoaka) patuxentia* in their species concept. All three of these taxa possess upturned shoulder ribs (projecting posteriorward) and this was presumed to be an important feature that could be used to unite the three completely unrelated species, regardless of number of ribs and the shell shapes. The upturned shoulder rib, however, is found in many other groups of ecphoras, especially in the genera *Ecphorosycon* and *Chesathais*, and cannot be used as a species-specific character. Ward and Gilinsky (1988: plate 2) illustrate two of their paratypes of *E. meganae*, both of which belong to different genera and represent separate species; one (paratype USNM 424300) is a *Trisecphora (Matoaka) patuxentia*, with three thin, undulating ribs (similar to the specimens shown here on Figure 4.15) and the other (paratype USNM 4243101) is a *Chesathais lindae*, which is almost identical to the specimens shown here on Figure 3.10. The type lot of *E. meganae* can now be seen to comprise a mixed lot, containing four separate species.

Ecphora amyae Petuch and Drolshagen, 2010
(Figure 5.5 A, B)

Description: (Reprinted from the original description; Petuch and Drolshagen, 2010:126-127, figure 5.4J, K) "Shell of average size for genus, cylindrical in shape, with high, protracted, scalariform spire, and sculptured with four thick, wide ribs; ribs flattened on edges, slightly "T"-shaped in cross section, ornamented with single large, deeply incised spiral thread, producing thin, shallow groove; shoulder rib extending upward (posteriorward), producing canaliculate spire whorls; subsutural areas of shoulder flat or slightly sloping; areas between ribs ornamented with numerous coarse spiral threads and strong longitudinal growth increments, giving shell rough, scaly appearance; umbilicus proportionally narrow; aperture oval in shape; siphonal canal ornamented with 15 to 17 large spiral cords."
Type Specimens: HOLOTYPE: length 37 mm, from Shattuck Zone 19 at Drumcliff, Jones Wharf, St. Mary's County, Maryland, along the Patuxent River, PRI 9918; PARATYPE: length 25 mm (incomplete), from the same locality and formation as the holotype, in the Petuch research collection.
Type Locality: Collected in the upper shell bed (Shattuck Zone 19) at Drumcliff, Jones Wharf, St. Mary's County, Maryland, along the Patuxent River; Boston Cliffs Member, Choptank Formation.
Stratigraphic Range: Confined to Shattuck Zone 19, Boston Cliffs Member, Choptank Formation of Maryland; late Serravallian Age of the middle Miocene.
Etymology: Named for Amy Plummer of Plum Point, Calvert County, Maryland.
Discussion: *Ecphora amyae* is a rare species, known from only a few incomplete specimens. By having slightly "T"-shaped ribs and by having an upwardly-extending shoulder rib, the new species is most similar to the Shattuck Zone 18 *E. meganae* and is the direct descendant of that Drumcliff Member ancestral species. The unrelated sympatric *E. (Powhatan) williamsi*, a member of the wide-ribbed subgenus *Powhatan*, was originally, and incorrectly, described as a subspecies of the older and stratigraphically-lower *E. meganae*. The discovery of *E. amyae* showed that *E. williamsi* is not the descendant of *E. meganae*, as proposed by Ward and Gilinsky (1988), but that it belongs to a separate lineage of wide-ribbed species that includes ancestral taxa such as *E. calvertensis* and *E. chesapeakensis*.

Ecphora kochi Ward and Gilinsky, 1988
(Figure 5.6 A, B, C, D)

Description: (abbreviated description, taken from Ward and Gilinsky, 1988: 10-11) "Shell large, globose, inflated with rounded body whorl; ribs four in number on body whorl, "T"-shaped in cross section, ornamented with one or two large, deeply impressed threads; ribs slightly undulating; subsutural area flat, producing stepped, scalariform spire; areas between ribs smooth, without spiral sculpture; umbilicus large, flaring."
Type Specimens: HOLOTYPE: length 89.6 mm, width 85.9 mm, below Cobham Wharf, Surry County, Virginia, on the James River, Cobham Bay Member of the Eastover Formation, early Messinian Age, late Miocene; USNM 405315; PARATYPES: (all from the same locality and formation as the holotype) length 82.5 mm, USNM 424308; length 89.0, USNM 424309; length 87.4 mm, USNM 424310.
Type Locality: In exposures of the Cobham Bay Member of the Eastover Formation, below Cobham Wharf, Surry County, Virginia, on the James River.
Stratigraphic Range: Confined to beds of late Messinian Age, late Miocene and found in the Cobham Bay Member, Eastover Formation of Virginia and North Carolina and in the lowest beds of the Goose Creek Formation in South Carolina.
Etymology: Named for Carl F. Koch, Department of Geophysical Sciences, Old Dominion University.
Discussion: Based on its stratigraphic position in late Messinian Miocene beds and also on its shell morphology, *E. kochi* can be seen to be the direct ancestor of the *E. quadricostata* species complex and represents the mid-point between the older *E. meganae*-type shells and the younger *E. quadricostata* complex.

Ecphora roxaneae Petuch, 1991
(Figure 5.7 A, B, C, D)

Description: (Reprinted from the original description; Petuch, 1991: 32-ee. plate 5, figures 15, 16) "Shell inflated, ovate, rotund, with rounded body whorl; spire elevated, scalariform; ribs four in number, with faint cordlike fifth rib around siphonal canal-body whorl juncture (as in holotype); some specimens with fifth rib fully developed and equal in size to four main ribs (as in paratype); ribs thick, rounded, "T"-shaped in cross section; surface of ribs sculptured with 3 to 5 incised grooves; areas between ribs smooth, with only faint spiral sculpture, most prominent on spire whorls; umbilicus wide, flaring; aperture wide and oval in shape."

Type Specimens: HOLOTYPE: length 67 mm, from Unit 11 in the APAC Pit, Sarasota, Sarasota County, Florida, CM 35622; PARATYPE: length 65 mm, from the basal bed in the DeSoto Shell Pit, Arcadia, DeSoto County, Florida, Sarasota Member, Murdock Formation, in the Petuch research collection.

Type Locality: From Unit 11, Sarasota Member, Murdock Formation, in the APAC Pit, Richardson Road, Sarasota, Sarasota County, Florida (now flooded as a lake in a county park).

Stratigraphic Range: Confined to beds of late Zanclean Age, early Pliocene and known only from quarries along southwestern Florida, in the Sarasota Member of the Murdock Formation.

Etymology: Named for Roxane Wilson of Arcadia, Florida.

Discussion: The Zanclean Pliocene *E. roxanae* is most similar to the older and stratigraphically-lower Messinian Age *E. kochi*, but differs in being a less inflated and more cylindrical shell with lower, more cordlike ribs that do not project from the body whorl.

Ecphora roxaneae dalli Petuch and Berschauer, new subspecies (Figure 5.8 A, B, C)

Description: Shell inflated, ovate, rotund, with rounded body whorl; spire elevated, scalariform; ribs four in number; ribs wide, "T"-shaped in cross section, projecting from the body whorl; surface of ribs sculptured with 3 to 5 incised grooves; areas between ribs smooth, with only faint raised spiral threads and tiny cords, most prominent on body whorl; umbilicus wide, flaring; aperture wide and oval in shape.
Type Specimens: HOLOTYPE: length 75.6 mm, width 63.5 mm, from the Martin Marietta Berkeley Quarry, Eutawville, Orangeburg County, South Carolina, Goose Creek Formation, PRI 104777; PARATYPES: length 70.4 mm, same locality and formation as the holotype, in the Berschauer research collection; length 111 mm, from the same locality and formation as the holotype, in the Petuch research collection.
Type Locality: Found embedded in a soft, vuggy limestone at the base of the Martin Marietta Berkeley Quarry, near Eutawville, Orangeburg County, South Carolina; in the basal bed of the Goose Creek Formation ("Goose Creek Limestone").
Stratigraphic Range: Confined to the lower Goose Creek Formation, Zanclean Age of the early Pliocene. Similar shells have been collected in the basal Yorktown Formation beds in the PotashCorps Phosphate Mine (old Lee Creek Phosphate Mine) at Aurora, North Carolina.
Etymology: Named for William Healey Dall, the late 19th-century and early 20th-century American malacologist, molluscan paleontologist, and geologist who contributed so much to our knowledge of the stratigraphy of the eastern United States.
Discussion: This large and distinctive shell from the Goose Creek Formation of South Carolina, is most similar to *Ecphora roxaneae* and is here considered to be a northern subspecies of the southern Floridian *E. roxanae*. The Goose Creek subspecies, *dalli*, differs primarily in having a more rotund and inflated shell and also in having the four body whorl ribs projecting farther from the body whorl than the lower and more adherent ribs seen on the nominate subspecies.

Ecphora leecreekensis Petuch, 1988
(Figure 5.9 A, B, C, D)

Description: (Reprinted from the original description, as *Ecphora (Ecphora) quadricostata leecreekensis*; Petuch, 1988: 131-132, plate A2, figure E) "Shell inflated, round, somewhat globose, subsutural area flattened, producing scalariform spire whorls; body whorl ornamented with 4 well developed, moderately wide ribs; ribs "T"-shaped in cross section; each rib ornamented with one, deeply-impressed groove; areas between ribs heavily sculptured with numerous raised spiral threads and small cords; holotype with large cord around anterior end of body whorl; siphonal canal well developed, elongated, heavily ornamented with coarse spiral threads; umbilicus narrow, poorly developed."

Type Specimens: HOLOTYPE: length 44 mm, from the lower Yorktown beds (Zone 1, Sunken Meadow Member) in the Lee Creek Phosphate Mine (now PotashCorps Phosphate Mine), Aurora, North Carolina, UF 20520; PARATYPE: length 42 mm, from the same locality and formation as the holotype, in the Petuch research collection.

Type Locality: From the Zone 1 beds of the Sunken Meadow Member, Yorktown Formation in the Lee Creek Phosphate Mine (now PotashCorps Phosphate Mine), Aurora, Beaufort County, North Carolina.

Stratigraphic Range: Confined to the lower beds (Sunken Meadow Member) of the Yorktown Formation in North Carolina; at present, known only from the phosphate mines at Aurora, North Carolina; early Zanclean Age, early Pliocene.

Etymology: Named for the old Lee Creek Phosphate Mine at Aurora, North Carolina, where the first specimens were collected and for the nearby Lee Creek.

Discussion: *Ecphora leecreekensis* is most similar to the younger and stratigraphically-higher *E. quadricostata*, but differs in being a slightly more globose shell with wider, heavier ribs, and in having coarse, threaded sculpture between the ribs and on the shoulder and subsutural area. The grooving on the ribs of *E. leecreekensis* is also deeper and wider than that found on *E. quadricostata*.

Ecphora quadricostata (Say, 1824)
(Figures 5.10 A, B, C, D; 5.11 A, B, C, D)

Description: (Reprinted from the original description, as *Fusus quadricostata*; Say, 1824: 127) "ovate ventricose; with a dilated umbilicus, and four much elevated belts, which are dilated at their tops. Spire short, the volutions with but two belts, the others being concealed by succeeding whorls; body whorl with four belts, which are equidistant, much elevated, wider at the top than at the junction with the whorl, and with one or two deeply impressed lines; intervening spaces wrinkled, the wrinkles extending over the belts; aperture suboval; canal short and contracted; labrum with a groove corresponding with each of the exterior ribs; umbilicus dilated, large, not visibly penetrating to the inner summit; the exterior margin prominent and deeply dentated."

Type Specimen: LECTOTYPE: length 82 mm, in the Finch Collection of the Natural History Museum, London, (British Museum (Natural History)), BM(NH) GG12661, London, United Kingdom. According to Ray (1987: 2), the original type of *Ecphora quadricostata* was housed in the Academy of Natural Sciences of Philadelphia, but is apparently lost and cannot be located. Wilson (1987: 22-23), although showing that this was a Pliocene species, never designated a neotype for the lost holotype. A lectotype was later designated, using an historical specimen in the Finch Collection of the British Museum, by Ward and Gilinsky (1988: 13).

Type Locality: Based upon the data associated with the lectotype in the Natural History Museum, London, the type of *Ecphora quadricostata* was collected along the York River, Virginia, from beds now referred to as Zone 2 (Rushmere Member) of the Yorktown Formation.

Stratigraphic Range: Confined to the earliest Piacenzian Age of the Pliocene and found in the Rushmere Member of the Yorktown Formation (lower Zone 2 Yorktown), the Goose Creek Formation of South Carolina, and the Jupiter Waterway Member of the Murdock Formation of southern Florida (possibly a stratigraphically-older unnamed subspecies).

Etymology: Named in reference to the shell having a sculpture composed of four prominent ribs.

Discussion: Martin (1904: 208) confused this Pliocene species with the large *Powhatan* species of the late Miocene, such as *Ecphora (Powhatan) gardnerae* and *E. (Powhatan) germonae* (the subgenus *Powhatan* is discussed and illustrated later in this chapter), erroneously listing *E. quadricostata* as a Maryland species. This error was later corrected by Wilson (1987) when he distinguished the true *E. quadricostata* from the well-known Maryland *Powhatan* species, showing that it was actually a much younger, Pliocene species from Virginia and North Carolina (also discussed in Campbell, 1993).

Ecphora quadricostata striatula Petuch, 1986
(Figure 5.12 A, B, C, D)

Description: (Reprinted from the original description, as *Ecphora striatula*; Petuch, 1996:406, plate 3, figures 15, 16) "Shell inflated, thin for genus, with four prominent ribs; ribs thin, bladelike, with edges flattened, ornamented with one or two deeply-incised spiral threads; rib along shoulder flat, horizontal in orientation; anterior three ribs tipped downward; areas between ribs and on spire ornamented with numerous conspicuous, fine raised cords and strong spiral threads; siphonal canal long, well developed, ornamented with numerous large, raised spiral cords; umbilicus open but narrow; outer calcitic layer of shell reddish-brown in color, with ribs being darker."

Type Specimens: HOLOTYPE: length 31 mm, width 29 mm, from Unit 10 in the APAC Pit, Sarasota, Sarasota County, Florida, Buckingham Member, Tamiami Formation, MCZ 29225; PARATYPE: fragment of spire, length 11 mm, from the Lakes of the Meadows construction excavations, SW147th and Bird Road, Miami, Florida, basal bed of the lower Golden Gate Member, Tamiami Formation (coral reef facies), MCZ 29235.

Type Locality: In Unit 10, Buckingham Member, Tamiami Formation, in the APAC Pit, Richardson Road, Sarasota, Sarasota County, Florida.

Stratigraphic Range: This ornately-sculptured subspecies is confined to beds of early Piacenzian Age of the Pliocene, and has been collected in equivalent units all across Florida, including the Buckingham Member of the Tamiami Formation at Sarasota, the lower part of the Golden Gate Member of the Tamiami Formation at Miami, and the lower part of the Jackson Bluff Formation at Alum Bluff (the "*Ecphora* Zone") along the Apalachicola River.

Etymology: Named in reference to the characteristic sculpture pattern composed of strong raised threads and small cords on the body and spire whorls.

Discussion: This early Piacenzian subspecies differs from its Zanclean ancestor, the nominate subspecies *Ecphora quadricostata quadricostata*, in having a far more ornate shell with numerous fine raised cords and spiral threads between the main ribs and in having thinner ribs that project farther from the body whorl. Further research may show that *E. quadricostata striatula* represents a full species, distinct from its older and stratigraphically-lower ancestor. For now, we consider it to be an early Piacenzian stratigraphic subspecies of the Zanclean *E. quadricostata*. This subspecies is also the farthest south-known ecphora, having been collected as recognizable fragments at the Bird Road housing construction site in Miami, Dade County, Florida.

Ecphora quadricostata umbilicata (Wagner, 1839)
(Figure 5.13 A, B, C, D)

Description: (not formerly described by Wagner, but redescribed by Martin, 1904: 209 and reprinted here) "This variety differs from typical forms of the species in having a thinner shell, a more flaring umbilicus, more loosely coiled whorls, and ribs that are not T-shaped. Because of the thinner, looser coiling, and the lithologic character of the formation in which it occurs, perfect specimens are far rarer than typical *quadricostata* and the distinctness of the variety has not been generally recognized. The figure prepared by Wagner in 1839 was on one of those plates which were not regularly published until 1898, but of which a few copies were distributed with manuscript names attached. Apparently no description of *var. umbilicata* has ever been published till now."
Type Specimen: HOLOTYPE: length 70 mm, from the Yorktown Formation of Virginia. The holotype was originally deposited in the Wagner Free Institute of Science in Philadelphia, but is now housed in the collections of the Academy of Natural Sciences of Drexel University, Philadelphia as an unnumbered specimen. The specimen shown here on Figure 5.13 A, B could be designated as a lectotype if the holotype cannot be found in the Academy collections.
Type Locality: No type locality was given originally; we here designate a new type locality based on a specimen collected recently and illustrated in Petuch, 2004 (plate 48, figures E, F); Rice's Pit, Hampton, Virginia, Morgarts Beach Member of the Yorktown Formation (upper Zone 2 Yorktown).
Stratigraphic Range: Confined to the late Piacenzian Age of the Pliocene and restricted to the Morgarts Beach Member of the Yorktown Formation of Virginia and the upper beds of the Pinecrest Member of the Tamiami Formation of southern Florida.
Etymology: Named in reference to the large flaring umbilicus that characterizes this species.
Discussion: Martin (1904: 209, plate LII, figure 4) lists this species as being present in the Serravallian Miocene Choptank Formation, but that has now been found to be in error. The Maryland Miocene shell that he illustrates (Martin, 1904: plate LII, figure 4, from "Jones Wharf, Maryland") was actually a previously-unnamed umbilicate species, later named *Planecphora vokesi*, and which is discussed in Chapter 6 of this book. The true *Ecphora quadricostata umbilicata* is a much younger species, being found in the Yorktown Formation of Virginia and North Carolina, the Duplin Formation of South Carolina, and the Pinecrest Member of the Tamiami Formation of Florida.

The senior author (Petuch, 1988:68-69) incorrectly synonymized *Ecphora quadricostata umbilicata* with *E. quadricostata*, citing Wilson's 1987 paper where he synonymized Wagner's taxon with Say's "*Fusus 4-costatus*." Subsequent study of more specimens of *E. quadricostata umbilicata* (two illustrated here in this book) demonstrates that the shell is morphologically-distinct from the nominate subspecies and occurs higher up in the stratigraphic record. Although the taxon was proposed and illustrated in 1839 (Wagner, plate 2, figure 2), a detailed formal description was not undertaken until 1904, when Martin (page 209) expanded upon Say's original description.

Ecphora quadricostata rachelae Petuch, 1988
(Figure 5.14 A, B, C, D)

Description: (Reprinted from the original description, as *Ecphora (Ecphora) quadricostata rachelae*; Petuch, 1988: 69-70, figure 33) "Shell smaller and more fusiform than nominate subspecies, with higher and distinctly sloping subsutural areas; body whorl with four wide, slightly rounded "T"-shaped ribs; ribs low and close to body whorl, being almost cordlike in form; edges of ribs ornamented with one or two impressed spiral threads; areas between ribs smooth, without sculpture; umbilicus narrow, poorly developed."
Type Specimens: HOLOTYPE: length 38 mm, from the Yadkin Pit, Chesapeake, Yadkin County, Virginia, in the Moore House Member, Yorktown Formation, UF 20512.
Type Locality: From an exposure of the Moore House Member, Yorktown Formation, in the Yadkin Pit, Chesapeake, Yadkin County, Virginia.
Stratigraphic Range: Confined to the late Piacenzian Age of the Pliocene and restricted to the Moore House Member of the Yorktown Formation (middle Yorktown Zone 2) and the middle beds of the Pinecrest Member of the Tamiami Formation of southern Florida. This high-spired subspecies has been collected in Moore House Member exposures at several localities, including: in the Yadkin Pit at Chesapeake, Yadkin County, Virginia; the Lone Star Cement Company Mine, Chuckatuck, Virginia; Rice's Pit, Hampton, Virginia: the PotashCorps Phosphate Mine (old Lee Creek Mine), Aurora, Beaufort County, North Carolina; in Unit 7 of the APAC Pit, Sarasota, Sarasota County, Florida; and in Unit 7 of the Quality Aggregates Pit #6, Sarasota, Sarasota County, Florida.
Etymology: Named for Rachel ("Raye") Germon, Department of Invertebrates, National Museum of Natural History, Smithsonian Institution.
Discussion: *Ecphora quadricostata rachelae* differs from the older, stratigraphically-lower nominate subspecies *E. quadricostata quadricostata* in being a smaller, more compact, and heavier shell with ribs that lie closer to the body whorl. The new subspecies is also much more fusiform than the nominate subspecies, having a higher spire, and more rounded shoulder. This late Piacenzian ecphora is the last-living species of *Ecphora s.s.* and disappears at the end of the Pinecrest Member, Tamiami Formation in Florida and at the end of Yorktown Formation deposition in Virginia and North Carolina.

Iconography of *Ecphora* sensu stricto Species

Figure 5.1 *Ecphora wardi* Petuch, 1988
A, B= paratype specimen, length 55.6 mm. Collected in Shattuck Zone 10 at Plum Point, Calvert County, Maryland; from the Plum Point Member, Calvert Formation; **C, D=** typical specimen, length 23 mm. Collected in the PotashCorps Mine, Aurora, Beaufort County, North Carolina; from the Bonnerton Member, Pungo River Formation.

Figure 5.2 ***Ecphora mattinglyi*** **Petuch, 2004**
A, B= paratype specimen, length 88 mm. Collected in Shattuck Zone 14, at Scientists Cliffs, Calvert County, Maryland; from the lower bed of the Calvert Beach Member, Calvert Formation.

Figure 5.3 ***Ecphora sandgatesensis*** **Petuch, 1988**
A, B= paratype specimen, length 40 mm; **C, D=** paratype specimen, length 44 mm. Both paratypes collected in Shattuck Zone 17 at Sandgates, Patuxent River, St. Mary's County, Maryland; from the St. Leonard Member, Choptank Formation.

Figure 5.4 *Ecphora meganae* Ward and Gilinsky, 1988
A, B= typical specimen, length 70 mm. Collected in Shattuck Zone 18 at Jones Wharf, Patuxent River, St, Mary's County, Maryland; from the Drumcliff Member, Choptank Formation; **C, D**= typical specimen, length 69 mm. Collected in Shattuck Zone 18 between Matoaka Cottages and Governor Run, St. Leonard, Calvert Cliffs, Calvert County, Maryland.

Figure 5.5 *Ecphora amyae* Petuch and Drolshagen, 2010
A, B= paratype specimen, length 25 mm (incomplete). Collected in Shattuck Zone 19 at Drumcliff, along the Patuxent River in St. Mary's County, Maryland; from the Boston Cliffs Member, Choptank Formation.

Figure 5.6 ***Ecphora kochi*** **Ward and Gilinsky, 1988**
A, B= typical specimen, length 92 mm, collected in the lower beds of the PotashCorps Mine, Aurora, Beaufort County, North Carolina; from the Cobham Bay Member, Eastover Formation.

Figure 5.7 *Ecphora roxaneae* Petuch, 1991
A, B= typical specimen, length 102 mm. Collected in Unit 11 in the APAC Pit, Sarasota, Sarasota County, Florida, Sarasota Member, Murdock Formation; **C, D=** paratype specimen, length 48 mm. Collected in the basal bed of the DeSoto Shell Pit, Arcadia, DeSoto County, Florida, Sarasota Member, Murdock Formation.

Figure 5.8 ***Ecphora roxaneae dalli*** **Petuch and Berschauer, new subspecies**
A= holotype, length 75.6 mm; **B, C=** apertural and dorsal views of paratype, length 70.4 mm. Both collected in the basal bed of the Martin Marietta Berkeley Quarry, Eutawville, Orangeburg County, South Carolina, Goose Creek Formation.

Figure 5.9 ***Ecphora leecreekensis*** **Petuch, 1988**
A, B= paratype specimen, length 42 mm; **C, D=** typical specimen, length 44 mm. Both specimens were collected in the basal section of the middle beds in the PotashCorps Mine, Aurora, Beaufort County, North Carolina; from the Sunken Meadow Member, Yorktown Formation (Yorktown Zone 1).

Figure 5.10 *Ecphora quadricostata* (Say, 1824). Specimens from Florida.
A, B= typical specimen, length 60 mm; **C, D**= specimen with undulating ribs, length 41 mm. Both specimens were collected in the basal bed in the APAC Pit, Sarasota, Sarasota County, Florida; from the Jupiter Waterway Member, Murdock Formation.

Figure 5.11 ***Ecphora quadricostata*** **(Say, 1824). Specimens from the Carolinas.**
A, B= typical specimen, length 44 mm; **C, D=** typical specimen, length 46 mm. Both specimens were collected in the middle beds of the PotashCorps Mine, Aurora, Beaufort County, North Carolina; from the Rushmere Member, Yorktown Formation; **E=** typical specimen, length 45 mm. Collected in the Martin Marietta Berkeley Quarry, east of Eutawville, Orangeburg, South Carolina; from the lower beds of the Goose Creek Formation.

Figure 5.12 *Ecphora quadricostata striatula* Petuch, 1986
A, B= typical specimen, length 49 mm. Collected in the middle bed at Alum Bluff, Liberty County, Florida, along the Apalachicola River; from the lower bed of the Jackson Bluff Formation; **C, D**= typical specimen with heavy spiral cords between the ribs, length 57 mm. Collected in Unit 10 in the Quality Aggregates Pit #6, Sarasota, Sarasota County, Florida; from the Buckingham Member, Tamiami Formation.

Figure 5.13 ***Ecphora quadricostata umbilicata*** **(Wagner, 1839)**
A, B= typical specimen, length 68 mm. Collected in Rice's Pit, Hampton, Virginia; from the Morgarts Beach Member, Yorktown Formation; **C, D=** typical specimen, length 96 mm. Collected in the basal section of Unit 7 in Quality Aggregates Pit #6, Sarasota, Sarasota County, Florida; from the basal bed of the Pinecrest Member, Tamiami Formation.

Figure 5.14 *Ecphora quadricostata rachelae* Petuch, 1988
A, B= typical specimen, length 42 mm. Collected in the uppermost section of the middle beds in the PotashCorps Mine, Aurora, Beaufort County, North Carolina; from the Moore House Member, Yorktown Formation; **C, D**= typical specimen, length 63 mm. Collected in the uppermost bed of Unit 7 in Quality Aggregates Pit #6, Sarasota, Sarasota County, Florida; from the upper bed of the Pinecrest Member, Tamiami Formation.

The Subgenus *Rhombecphora* Petuch and Berschauer, new subgenus

SYSTEMATICS
Class Gastropoda Cuvier, 1797
Subclass Orthogastropoda Ponder and Lindberg, 1995
Superorder Caenogastropoda Cox, 1960
Order Sorbeoconcha Ponder and Lindberg, 1997
Infraorder Neogastropoda Wenz, 1938
Superfamily Muricoidea Rafineasque, 1815
Family Muricidae Rafinesque, 1815
Subfamily Ocenebrinae Cossmann, 1903
Tribe Ecphorini Petuch and Berschauer, new tribe
Genus *Ecphora* Conrad, 1843
Subgenus *Rhombecphora* Petuch and Berschauer, new subgenus

Diagnosis: Shells of average size for Tribe, fusiform, elongated, with distinctly rhomboidal body whorls and shell outlines; spires high and protracted; shoulder and subsutural areas distinctly sloping; shells ornamented with 4 thin, low, adherent spiral cordlike ribs; ribs sculptured with single large, deeply-impressed spiral thread; shoulder rib of equal size to other 3 ribs, carinate, not projecting posteriorward; areas between ribs ornamented with very faint spiral threads and tiny cords and very numerous longitudinal growth lines, giving shells squamose appearance; umbilicus narrow; apertures oval in shape.
Type Species: *Ecphora (Rhombecphora) rikeri* Petuch, 1988, Shattuck Zone 18, Drumcliff Member, Choptank Formation of Maryland, Serravallian Age, middle Miocene.
Species in *Rhombecphora*: *Ecphora (Rhombecphora) harasewychi* Petuch, 1988, Shattuck Zone 17, St. Leonard Member, Choptank Formation; *Ecphora (Rhombecphora) rikeri* Petuch, 1988, Shattuck Zone 18, Drumcliff Member, Choptank Formation.
Etymology: Named for the distinct rhomboidal shell shape and shell outline.
Discussion: The new subgenus represents a small and short-lived clade that broke off from the main *Ecphora s.s.* lineage at the beginning of the Serravallian Age, possibly utilizing new ecological resources that became available during the cold climate interval of the middle Miocene. Unlike the contemporaneous *Ecphora sandgatesensis* (Shattuck Zone 17) and *E. meganae* (Shattuck Zone 18), *Rhombecphora* species lack the distinctly upward-pointing, bladelike shoulder rib that characterizes those early *Ecphora s.s.* species. By the end of Shattuck Zone 18 time, this small clade became extinct and is not present in Shattuck Zone 19 (Boston Cliffs Member) nor in the subsequent St. Mary's Formation.

Ecphora (Rhombecphora) harasewychi **Petuch, 1988**
(Figure 5.15 A, B)

Description: (Reprinted from the original description; as *Ecphora (Ecphora) rikeri harasewychi*; Petuch, 1988e: 132, plate A2, figure A) "Shell fusiform, with rounded body whorl profile; spire high, protracted; body whorl ornamented with 4 low, cordlike ribs; ribs rounded in cross section; areas between ribs ornamented with numerous faint, fine, raised threads; faint, poorly developed fifth rib present around anterior end of body whorl; siphonal canal elongated, well developed; umbilicus open, well developed; subsutural and shoulder area sloping."

Type Specimens: HOLOTYPE: length 38 mm (incomplete), St. Leonard Member (Shattuck Zone 17) of the Choptank Formation at Sandgates, St. Mary's County, Maryland, UF 20519; PARATYPE: length 32 mm, same locality and formation as the holotype, in the Petuch research collection.

Type Locality: In a low cliff, along the waterline of the Patuxent River at Sandgates, St. Mary's County, Maryland, in Shattuck Zone 17 of the St. Leonard Member, Choptank Formation.

Stratigraphic Range: Confined to Shattuck Zone 17 of the St. Leonard Member, Choptank Formation, early Serravallian Age of the early middle Miocene. Known only from Zone 17 beds along the Patuxent River, where eroded and broken specimens are frequently found exposed at low tide on mud flats adjacent to the shoreline cliffs.

Etymology: Named for Dr. M.G. Harasewych of the Division of Mollusks, National Museum of Natural History, Smithsonian Institution.

Discussion: *Ecphora (Rhombecphora) harasewychi* from Shattuck Zone 17 is most similar to *E. (Rhombecphora) rikeri* (described next), the only other known member of the subgenus *Rhombecphora*, in being a smaller shell with better developed ribs, by having a less elongated and more compact appearance, and in having fewer and coarser incised spiral threads between the ribs. Like all the ecphoras found in the highly-leached Shattuck Zone 17, *E. (Rhombecphora) harasewychi* is always poorly preserved and almost always found in a cracked and fragmentary state.

Ecphora (Rhombecphora) rikeri Petuch, 1988
(Figure 5.16 A, B)

Description: (Reprinted from the original description: Petuch, 1988; 11-12) "Shell elongated, fusiform, with high protracted spire and 4 ribs; ribs low, cordlike, attached to body whorl; surface of cords bisected by single, deeply-impressed line; body whorl between ribs ornamented with numerous fine, raised spiral threads; siphonal canal well-developed; umbilicus open, well-developed."
Type Specimens: HOLOTYPE: length 54 mm, width 43 mm, Governor Run Pier, Calvert Cliffs, Maryland, Shattuck Zone 18, Drumcliff Member, Choptank Formation, USNM 424250; PARATYPES: length 33 mm, Shattuck Zone 18 at Jones Wharf, Drumcliff, St. Mary's County, Maryland, USNM 424251; length 80 mm, in Shattuck Zone 18 at Governor Run Pier, Calvert Cliffs, Calvert County, Maryland, in the research collection of the senior author.
Type Locality: In Shattuck Zone 18, Drumcliff Member of the Choptank Formation, at Governor Run, Calvert County, Maryland.
Stratigraphic Range: Confined to Shattuck Zone 18, Drumcliff Member, Choptank Formation, Serravallian Age of the Miocene.
Etymology: Named for Norman Riker, who collected the holotype at Governor Run.
Discussion: This distinctive ecphora differs from the older and stratigraphically-lower *Ecphora (Rhombecphora) harasewychi* from the St. Leonards Member in being a much larger shell with a narrower body whorl and in having a higher, more protracted spire. The ribs of *E. (Rhombecphora) rikeri* are also proportionally thinner and more cordlike than those seen on the older *E. (Rhombecphora) harasewychi*.

Iconography of *Rhombecphora* Species

Figure 5.15 *Ecphora (Rhombecphora) harasewychi* Petuch, 1989
A, B= typical specimen, length 32 mm. Collected in Shattuck Zone 17 at Sandgates, near Queentree, along the Patuxent River, St Mary's County, Maryland; from the St. Leonard Member, Choptank Formation.

Figure 5.16 *Ecphora (Rhombecphora) rikeri* Petuch, 1988
A, B= typical specimen, length 80 mm. Collected in Shattuck Zone 18 between Matoaka Cottages and Governor Run, St. Leonard, Calvert Cliffs, Calvert County, Maryland; from the Drumcliff Member, Choptank Formation.

The Subgenus *Powhatan* Petuch and Berschauer, new subgenus

The largest species of the entire Tribe Ecphorini belong to the subgenus *Powhatan*, including some very large, grapefruit-sized taxa such as *E. (Powhatan) asheri*, *E. (Powhatan) gardnerae*, *E. (Powhatan) whiteoakensis*. The ribs of *Powhatan* species are wider than those of true *Ecphora s.s.*, which are characteristically thinner and narrower. The body whorls of *Powhatan* species, especially in the more advanced late Miocene forms, are also much more inflated and globose than those of *Ecphora s.s.* species.

SYSTEMATICS
Class Gastropoda Cuvier, 1797
Subclass Orthogastropoda Ponder and Lindberg, 1995
Superorder Caenogastropoda Cox, 1960
Order Sorbeoconcha Ponder and Lindberg, 1997
Infraorder Neogastropoda Wenz, 1938
Superfamily Muricoidea Rafinesque, 1815
Family Muricidae Rafinesque, 1815
Subfamily Ocenebrinae Cossmann, 1903
Tribe Ecphorini Petuch and Berschauer, new tribe
Genus *Ecphora* Conrad, 1843
Subgenus *Powhatan* Petuch and Berschauer, new subgenus

Diagnosis: Shells largest of genus *Ecphora*, especially in geologically-youngest species, inflated, globose, heavy; shells with 4 or 5 large, wide, prominent ribs that are distinctly "T"-shaped in cross section; wide ribs frequently ornamented with numerous incised spiral threads; apertures very flaring and open, roughly circular in size.
Type Species: *Ecphora (Powhatan) asheri* Petuch, 1988, Shattuck Zone 21 of the Little Cove Point Member, St. Mary's Formation, early Tortonian Age, late Miocene.
Species in *Powhatan*: *Ecphora (Powhatan) calvertensis* Petuch, 1988, Calvert Beach Member (Shattuck Zone 14), Calvert Formation; *Ecphora (Powhatan) chesapeakensis* Petuch, 1988, Calvert Beach Formation (Shattuck Zone 16), Calvert Formation; *Ecphora (Powhatan) williamsi* Ward and Gilinsky, 1988, Boston Cliffs Member (Shattuck Zone 19), Choptank Formation; *Ecphora (Powhatan) conoyensis* Petuch, 2004, Conoy Member (Shattuck Zone 20), St. Mary's Formation; *Ecphora (Powhatan) asheri* Petuch, 1988, Little Cove Point Member (Shattuck Zone 21), St. Mary's Formation; *Ecphora (Powhatan) germonae* Ward and Gilinsky, 1988, Little Cove Point Member (Shattuck Zone 22), St. Mary's Formation; *Ecphora (Powhatan) gardnerae* Wilson, 1987, Windmill Point Member (Shattuck Zone 24), St. Mary's Formation; *Ecphora (Powhatan) gardnerae angusticostata* Petuch, 1988, Chancellor Point Sandstone ("Shattuck Zone 25"), St. Mary's Formation; *Ecphora (Powhatan) whiteoakensis* Ward and Gilinsky, 1988, Claremont Manor Member, Eastover Formation; *Ecphora (Powhatan) whiteoakensis hopei* Petuch, 1991, Cocoplum Member, Bayshore Formation; *Ecphora*

(Powhatan) pachycostata Petuch, 1988, Sunken Meadow Member, Yorktown Formation and Jupiter Waterway Member, Murdock Formation.
Etymology: Named for Powhatan, Paramount Chief of the Tsenacommacah of Chesapeake Bay and father of Princess Matoaka (Pocahontas). The name translates from Eastern Algonquian directly to "Paramount Chief", in reference to the impressive size of this group of ecphoras.

Ecphora (Powhatan) cf. *whiteoakensis* Ward and Gilinsky, 1988, length 83 mm. From an exposure of the Eastover Formation in the Lee Creek Phosphate Mine (now PotashCorps Phosphate Mine) at Aurora, Beaufort County, North Carolina. This high-ribbed variant may represent an unnamed species or subspecies and is part of a large complex of *Powhatan* taxa that evolved during the late Tortonian and Messinian Miocene.

Ecphora (Powhatan) calvertensis Petuch, 1988
(Figure 5.17 A, B)

Description: (Reprinted from the original description: Petuch, 1988; 64-65, figures 26 and 27) "Shell cylindrical, barrel-shaped, with 4 broad ribs; fourth rib (anteriormost) somewhat variable in width, with some specimens being narrower than other 3 ribs; ribs "T"-shaped in cross section; siphonal canal long and well-developed, sculptured with fine raised cords; ribs ornamented with 3-4 incised spiral threads; umbilicus small, narrow."
Type Specimens: HOLOTYPE: length 45 mm, width 36 mm, Shattuck Zone 14 at Scientist's Cliffs, Calvert Cliffs, Calvert County, Maryland, USNM 424256; PARATYPES: length 40 mm, same locality and stratigraphic unit as the holotype, USNM 424257; length 43 mm, same locality and stratigraphic unit as the holotype, in the research collection of the senior author.
Type Locality: Shattuck Zone 14 at Scientist's Cliffs, Calvert Cliffs, Calvert County, Maryland, Calvert Beach Member of the Calvert Formation.
Stratigraphic Range: Confined to Shattuck Zone 14 of the Calvert Beach Member, Calvert Formation, Langhian Age, Miocene.
Etymology: Named for the Calvert Formation.
Discussion: This is the oldest species in the subgenus *Powhatan* and the first four-ribbed species to have developed wide, extremely "T"-shaped ribs. These wide ribs persisted throughout the evolutionary history of the subgenus, from the Langhian Age of the Miocene to the Zanclean Age of the Pliocene. Small, fragmentary, and badly-preserved specimens of a possible new subspecies of *Ecphora (Powhatan) calvertensis* have been found in Shattuck Zone 11 (Plum Point Member) of the Calvert Formation south of Chesapeake Beach, Calvert County.

Ecphora (Powhatan) chesapeakensis Petuch, 1988
(Figure 5.18 A, B)

Description: (Reprinted from the original description, as *Ecphora (Ecphora) chesapeakensis*; Petuch, 1988: 68-69, figures 3, 4) "Shell cylindrical in shape, slightly inflated; shoulder rounded; body whorl ornamented with 4 large, wide, rounded, adherent cordlike ribs that are slightly "T"-shaped in cross section; ribs sculptured with 1-4 thin, impressed spiral threads; shoulder rib largest and widest, curving upward (posteriorward) to produce canaliculated spire whorls; wide shoulder rib slightly incurved, producing distinctly rounded appearance; areas between ribs relatively smooth, with only few very fine spiral threads; siphonal canal well developed, ornamented with numerous large spiral threads; umbilicus narrow."

Type Specimens: HOLOTYPE: length (incomplete) 28 mm, in Shattuck Zone 16, Calvert Beach Member, Calvert Formation, north of Scientists Cliffs, Calvert County, Maryland, UF 23798; PARATYPE: length 41 mm, same locality and formation as the holotype, in the Petuch research collection.

Type Locality: In exposures of Shattuck Zone 16, Calvert Beach Member of the Calvert Formation, 0.5 km north of Scientists Cliffs State Park, Calvert Cliffs, Calvert County, Maryland.

Stratigraphic Range: Known only from Shattuck Zone 16, Calvert Beach Member, Calvert Formation, late Langhian Age of the early middle Miocene.

Etymology: Named for Chesapeake Bay, which borders the type locality.

Discussion: *Ecphora (Powahatan) chesapeakensis* most closely resembles the younger and stratigraphically-higher *E. (Powhatan) williamsi* from Shattuck Zone 19 of the Boston Cliffs Member of the Choptank Formation (discussed next). The Zone 16 species differs from its Zone 19 descendant in being a smaller, less inflated shell with thinner, less rounded ribs that are less adherent and project farther from the body whorl. Both species have similar wide, rounded, incurved shoulder ribs.

Ecphora (Powhatan) williamsi Ward and Gilinsky, 1988
(Figure 5.19 A, B)

Description: (abbreviated description taken from Ward and Gilinsky, 1988: 5-6; as *Ecphora meganae williamsi*) "Shell cylindrical with rounded shoulder, solid, thickened; body whorl ornamented with four wide, thick, rounded ribs; smaller, thinner fifth rib around anterior of body whorl sometimes present; entire shell and ribs heavily sculptured with numerous fine raised threads and small cords; umbilicus wide, flaring; large secondary cords sometimes present between ribs."

Type Specimens: HOLOTYPE: length 55.4 mm, width 48 mm, western side of Chesapeake Bay, near Flag Ponds, Calvert County, Maryland, USNM 495319; PARATYPES: (all from the same locality and member as the holotype) length 58.5 mm, USNM 424304; length 40.7 mm, USNM 424305.

Type Locality: 1.0 mile below Flag Ponds, on the western side of Chesapeake Bay, Calvert County, Maryland, in the Boston Cliffs Member (Shattuck Zone 19) of the Choptank Formation.

Stratigraphic Range: Confined to the Boston Cliffs Member (Shattuck Zone 19), Choptank Formation of Maryland. Found in Shattuck Zone 19 exposures along the Calvert Cliffs, Calvert County, Maryland and also in the upper shell bed (Shattuck Zone 19) at Drumcliff, Jones Wharf, St. Mary's County, Maryland, along the Patuxent River; late Serravallian Age, middle Miocene.

Etymology: Named for S. Jeffress Williams, U.S. Geological Survey.

Discussion: Although being structurally and morphologically very different, *E. (Powhatan) williamsi* was originally described as a subspecies of the very dissimilar *Ecphora meganae* (discussed earlier in this chapter).

Ecphora (Powhatan) conoyensis Petuch, 2004
(Figure 5.20 A, B)

Description: (Reprinted from the original description, as *Ecphora conoyensis* Petuch, 2004: 281-282, plate 31 D, E) "Shell small for genus, thick and compact; whorls rounded; spire proportionally low; body whorl ornamented with 4 large, wide, rounded ribs; fifth low, faint rib present around base of siphonal canal; large spiral threads and tiny cords present between ribs and between anteriormost rib and faint fifth rib; spire whorls with 2 rounded ribs and numerous spiral threads; siphonal canal narrow and elongated; umbilicus proportionally narrow; aperture wide and flaring, almost circular; interior of aperture ornamented with 12-15 large, prominent raised cords."

Type Specimens: HOLOTYPE: length 52 mm, width 43 mm, from Camp Conoy, Calvert County, Maryland, Conoy Member, St. Mary's Formation, AMNH FI-50667; PARATYPE: length 47 mm, from the same locality and formation as the holotype, in the Petuch research collection.

Type Locality: From Shattuck Zone 20, exposed in the cliffs at Camp Conoy, near Calvert Cliffs State Park, north of Cove Point, Calvert County, Maryland.

Stratigraphic Range: Confined to the Conoy Member (Shattuck Zone 20) of the St. Mary's Formation, along the southern Calvert Cliffs of Calvert County, Maryland; latest Serravallian Age of the late middle Miocene.

Etymology: Named for Camp Conoy, Calvert County, Maryland, the type locality.

Discussion: *Ecphora (Powhatan) conoyensis* represents the morphological intermediate between the older, stratigraphically-lower *E. (Powhatan) williamsi* from Shattuck Zone 19 of the Choptank Formation and the younger, stratigraphically-higher *E. (Powhatan) asheri* from Shattuck Zone 21 of the St. Mary's Formation. The Zone 19 species retains the small size, stocky thick shell, and narrow umbilicus of *E. williamsi* but has the flattened ribs of *E. asheri*. The small, faint fifth rib of *E. conoyensis* presages the large, equal-sized fifth rib of *E. asheri*.

Ecphora (Powhatan) asheri Petuch, 1988
(Figure 5.21 A, B)

Description: (Reprinted from the original description; Petuch, 1988: 91-92, figure 41) "Shell very globose, rounded, inflated, with 5 large ribs; ribs broad, "T"-shaped in cross section; spire whorls with 3 ribs; thin, thread-like rib between 5 main ribs on some specimens; umbilicus well-developed, flaring."
Type Specimens: HOLOTYPE: length 55 mm, width 48 mm, USNM 424246; PARATYPES: length 82 mm, USNM 4244247; length 75 mm, in the research collection of the senior author.
Type Locality: Collected in Shattuck Zone 21 at the Chesapeake Ranch Club, Little Cove Point, Calvert County, Maryland; from the lower bed of the Little Cove Point Member, St. Mary's Formation.
Stratigraphic Range: Confined to Shattuck Zone 21 of the Little Cove Point Member of the St. Mary's Formation, early Tortonian Age of the Middle Miocene. This large, distinctive, 5-ribbed species is narrowly confined stratigraphically to Zone 21 and is known only from the area of the type locality near the Chesapeake Ranch Club north of Solomons, Calvert County, Maryland.
Etymology: Named for Donald Asher of Mechanicsville, Maryland, who aided the senior author in his exploration of the Calvert Cliffs fossil beds.
Discussion: This distinctive muricid represents the first appearance of a very large, wide-ribbed *Powhatan* in the fossil record, foreshadowing the other large species of the late Miocene and early Pliocene. *Ecphora (Powhatan) asheri* is most similar to *E. (Powhatan) germonae* from Shattuck Zone 22 of the St. Mary's Formation (described next) but differs in being a more rounded and inflated shell with ribs that are more adherent, flattened, and closer to the body whorl. Unlike *E. (Powhatan) germonae* and the other subsequent large ecphoras, *E. (Powhatan) asheri* characteristically has five equal-sized ribs instead of four, and is a more ornately sculptured species with smaller cords being present between the five main ribs. These vary in number from two to five, but most often occur in sets of three, like the specimen illustrated here (Figure 5.22 A, B). Some extreme individuals, which have been found as fragmentary specimens, have smaller ribs between the main ribs and also numerous strong intercostal cords, producing shells with as many as 10 ribs and 12 cords.

Ecphora (Powhatan) germonae Ward and Gilinski, 1988
(Figure 5.22 A, B, C, D)

Description: (abbreviated description taken from Ward and Gilinsky, 1988: 6-7; as *Ecphora gardnerae germonae*) "Shell large, globose in shape, with flattened subsutural area; body whorl ornamented with four large, wide ribs that are "T"-shaped in cross section; some specimens have faint, nearly obsolete fifth rib, which is cordlike in form, around the anterior of the body whorl; areas without spiral sculpture; umbilicus well developed, slightly flaring."
Type Specimens: HOLOTYPE: length 78 mm, width 67.8 mm, in Shattuck Zone 22 at Little Cove Point, Calvert County, Maryland, USNM 405318; PARATYPES: length 66.9 mm, USNM 424306 and length 66 mm, USNM 424307, both from near the Chesapeake Ranch Club, Little Cove Point, Calvert County, Maryland; Little Cove Point Member (Shattuck Zone 22), St. Mary's Formation, early Tortonian Age of the late Miocene.
Type Locality: 0.5 mile south of Little Cove Point, Calvert County, Maryland in 3 foot thick shell bed, 20 feet above mean sea level; Little Cove Point Member (Shattuck Zone 22), St. Mary's Formation.
Stratigraphic Range: Confined to Shattuck Zone 22, Little Cove Point Member of the St. Mary's Formation, Calvert County, Maryland, early Tortonian Age of the late Miocene. Found only in the vicinity of Little Cove Point and the Chesapeake Ranch Club, Calvert County.
Etymology: Named for Rachel ("Raye") Germon, Department of Invertebrates, National Museum of Natural History, Smithsonian Institution.
Discussion: *Ecphora (Powhatan) germonae* is the morphological intermediate between the older *E. (Powhatan) asheri* (Shattuck Zone 21) and the younger *E. (Powhatan) gardnerae* (Shattuck Zone 24); being much less globose than *asheri* but being more cylindrical than *gardnerae.* This is the only ecphora found in Shattuck Zone 22 and it is quite common in the sandy clays of the Little Cove Point Member near Cove Point, Calvert County, Maryland. Because of its thick shell and good preservation, *E. (Powhatan) germonae* is often collected as beach specimens at Little Cove Point, where it weathers out of the high cliff face and tumbles down to the high tide line without significant shell damage.

Ecphora (Powhatan) gardnerae Wilson, 1987
(Figures 5.23 A, B, C, D; 5.24 A, B)

Description: (condensed from the original description) Shell large for genus, inflated, with rounded body whorl; subsutural area flat, shoulder angled; body whorl ornamented with four large, wide ribs that are distinctly "T"-shaped in cross section; ribs elevated, projecting from body whorl; areas between ribs smooth, without spiral sculpture; siphonal canal very elongated; umbilicus well developed but narrow.
Type Specimen: HOLOTYPE: length 120 mm, from the St. Mary's River, St. Mary's County, Maryland, Windmill Point Member (Shattuck Zone 24), St. Mary's Formation, USNM 647519.
Type Locality: From Shattuck Zone 24, Windmill Point Member, St. Mary's Formation, along the St. Mary's River, St. Mary's County, Maryland. Since Wilson (1987) never actually designated a specific type locality, we here designate Shattuck Zone 24 at Chancellor Point, near St. Mary's City, St. Mary's County, Maryland, as the type locality.
Stratigraphic Range: Confined to Shattuck Zone 24 of the Windmill Point Member, St. Mary's Formation, early Tortonian Age of the late Miocene. The species has been collected mostly along the Chancellor Point and Windmill Point areas of the St. Mary's River, Maryland.
Etymology: Named for Dr. Julia Gardner of the Smithsonian Institution and U.S. Geological Survey, in recognition of her many important contributions to molluscan paleontology.
Discussion: *Ecphora (Powhatan) gardnerae* is one of the largest species of ecphoras, often reaching sizes of over 120 mm in length. The high, elevated, "T"-shaped ribs of this large shell are the best-developed in the subgenus *Powhatan*, projecting farther from the body whorl and spire whorl than the older and stratigraphically-lower *E. (Powhatan) germonae*. This beautiful and large St. Mary's ecphora is the Maryland State Shell.

Ecphora (Powhatan) gardnerae angusticostata **Petuch, 1988**
(Figure 5.25 A, B)

Description: (reprinted from the original description, as *Ecphora (Ecphora) gardnerae angusticostata*; Petuch, 1988: 93-94, figure 44) "Shell smaller and more delicate than nominate subspecies, with cylindrical, less globose outline; shoulder sharply angled, subsutural areas flattened; ribs thin, only slightly "T"-shaped in cross section; area between ribs smooth, without spiral sculpture; shoulder rib slightly upturned; umbilicus elongated, narrow."

Type Specimens: HOLOTYPE: length 38 mm, from indurated sandstone layer at top of Shattuck Zone 24, Windmill Point Member, St. Mary's Formation, St. Mary's River, Maryland, UF 20515; PARATYPE: length 40 mm, same locality and formation as the holotype, in the Petuch research collection.

Type Locality: Collected in a hard sandstone ledge cropping out suprajacent to Shattuck Zone 24, .5 km east of Chancellor Point, St. Mary's River, St. Mary's County, Maryland; from the Chancellor Point Sandstone ("Shattuck Zone 25"), St. Mary's Formation.

Stratigraphic Range: Confined to the thin sandstone layer of Shattuck Zone 25, Chancellor Point Sandstone, St. Mary's Formation, Tortonian Age, late Miocene. Known only from the Chancellor Point Sandstone.

Etymology: A combination of "angustus", Latin for "narrow", and "costata", meaning "ribbed"; in reference to the shell's characteristic narrow ribs.

Discussion: The Chancellor Point Sandstone *Ecphora (Powhatan) gardnerae angusticostata* differs from the older, stratigraphically-lower nominate subspecies in being a much smaller, more cylindrical shell with thinner, less "T"-shaped ribs. This small, delicate subspecies of *E. gardnerae* has only been found within the 40 cm-50 cm thick shelf of fossiliferous sandstone layer, which is exposed on a stretch of the St. Mary's River from Chancellor Point to St. Inigoes Creek. The sandstone unit, which is separated by a distinct unconformity from the underlying fossiliferous blue clays of Shattuck Zone 24, is a much younger unit and contains a completely different molluscan fauna than the subjacent Windmill Point Member (see Petuch and Drolshagen, 2010: 91-99 for more information on the Chancellor Point Sandstone and its unique molluscan fauna).

Ecphora (Powhatan) grulkei Petuch and Berschauer, new species (Figure 5.26 A, B, C)

Description: Shell small for subgenus, elongated and fusiform, broadly cylindrical in shape; spire stepped and scalariform; body whorl ornamented with four very wide, nearly-adherent and flattened ribs; areas between ribs ornamented with small, faint spiral threads; siphonal canal proportionally very long and narrow.
Type Specimens: HOLOTYPE: length 76 mm, from the base of the Claremont Manor Member of the Eastover Formation in the PotashCorps Phosphate Mine, Aurora, North Carolina, PRI 104778; PARATYPE: length 68 mm (fragmentary), from the same locality and formation as the holotype, in the Petuch research collection.
Type Locality: Collected in the lower section of the PotashCorps Mine (old Lee Creek Phosphate Mine), Aurora, Beaufort County, North Carolina; from the basal bed of the Claremont Manor Member, Eastover Formation, directly suprajacent to the bedding plane surface of the underlying Pungo River Formation.
Stratigraphic Range: Confined to the basal bed of the Claremont Manor Member, Eastover Formation, early Messinian Age of the late Miocene. At present, known only from the lowest Eastover beds in the Aurora, North Carolina phosphate mines. The type specimens were found embedded in a dark khaki-green sandy clay that was in direct contact with the limestone layer of the Bonnerton Member of the Pungo River Formation.
Etymology: Named for Wolfgang Grulke of Oborne, Dorset, England, in recognition of his important contributions to molluscan paleontology, especially the fossil cephalopods.
Discussion: *Ecphora (Powhatan) grulkei* is most similar to the younger and stratigraphically-higher *E. (Powhatan) whiteoakensis* (discussed next) from the upper beds of the Claremont Manor Member, but differs in having a much smaller and more cylindrical shell and in having proportionally wider and lower, almost cordlike, ribs. On the last part of the body whorl, the ribs flatten even more, becoming almost adherent to the body whorl.

Ecphora (Powhatan) whiteoakensis Ward and Gilinski, 1988
(Figure 5.27 A, B, C, D)

Description: (abbreviated description taken from Ward and Gilinsky, 1988:9-10; originally described as *Ecphora gardnerae whiteoakensis*) "Shell large, globose, inflated, with rounded outline; subsutural areas flat, with stepped spire; body whorl with four wide, slightly rounded ribs that are low and nearly adherent; areas between ribs smooth, lacking spiral sculpture; some specimens with fifth, obsolete, very faint cordlike rib around the anterior of the body whorl; umbilicus wide and flaring; ribs ornamented with 1-4 impressed spiral threads."

Type Specimens: HOLOTYPE: length 70.2 mm, width 66.5 mm, from White Oak Landing on the Mattaponi River, Virginia, Claremont Member of the Eastover Formation, late Tortonian Miocene; USNM 380706; PARATYPE: length 72 mm, width 64 mm, from the same locality and formation as the holotype, USNM 405316.

Type Locality: In exposures of the Claremont Manor Member, Eastover Formation, along the Mattaponi River at White Oak Landing, King William County, Virginia.

Stratigraphic Range: Confined to the lower beds of the Eastover Formation (Claremont Member) in Virginia and North Carolina, late Tortonian Age of the late Miocene. This species has also been collected from Eastover Formation exposures in the phosphate mines at Aurora, Beaufort County, North Carolina (Figure 5.27 C, D).

Etymology: Named for White Oak Landing on the Mattaponi River, the type locality.

Discussion: The morphological differences between this Eastover Formation species and the closely-related *Ecphora (Powhatan) whiteoakensis hopei* from the Bayshore Formation are outlined in the discussion at the end of the next species description.

Ecphora (Powhatan) whiteoakensis hopei Petuch, 1991
(Figure 5.28 A, B)

Description: (Reprinted from the original description, as *Ecphora (Ecphora) hopei*; Petuch, 1991:32, plate 5, figures 15, 16) "Shell inflated, globose, with rounded outline; subsutural area flat; spire stepped; body whorl with four very wide, slightly rounded ribs that are low and nearly adherent; fifth faint rib present around body whorl-siphonal canal juncture; areas between ribs ornamented with very faint spiral threads; surface of ribs sculptured with 2-3 incised threads; umbilicus wide and flaring; aperture oval in shape."
Type Specimens: HOLOTYPE: length 66 mm, from a construction pit at Murdock, Charlotte County, Florida, Cocoplum Member of the Bayshore Formation, CM 35623; PARATYPE: length 43 mm, same locality and formation as the holotype, CM 35624.
Type Locality: Dredged from a 3 m deep trench in a housing construction pit along Toledo Blade Boulevard, Murdock, Charlotte County, Florida.
Stratigraphic Range: Confined to beds of late Messinian Age, late Miocene of southwestern Florida; in the Cocoplum Member of the Bayshore Formation. Known only from the lower beds of the Bayshore Formation.
Etymology: Named for Mitchell Hope of Englewood, Florida.
Discussion: *Ecphora (Powhatan) whiteoakensis hopei* is most similar to the nominate subspecies, *E. (Powhatan) whiteoakensis whiteoakensis*, from the contemporaneous Eastover Formation of Virginia (Petuch and Berschauer, 2022: 53), but differs in being a more inflated shell with proportionally wider ribs that consistently project farther from the body whorl. Besides having a different rib structure, *E. (Powhatan) whiteoakensis hopei* frequently has thin threadlike cords present between the four main ribs, approximating the ornate multi-ribbed appearance of the St. Mary's Formation *E. (Powhatan) asheri*. Most specimens of this subspecies also have a well-developed, but smaller, fifth rib present around the body whorl-siphonal canal juncture. Although considered here to be a southern subspecies of the Eastover Formation nominate species, further research may show that *E. (Powhatan) whiteoakensis hopei* is a separate, distinct species.

Ecphora (Powhatan) pachycostata Petuch, 1988
(Figure 5.29 A, B, C, D)

Description: (Reprinted from the original description, as *Ecphora (Ecphora) pachycostata*, with a few emendations; Petuch, 1988: 130-131, plate A1, figure D) "Shell inflated, ovate, with rounded body whorl outline; spire moderately protracted; shoulder and subsutural area flattened; body whorl ornamented with four very wide, very thickened rib; ribs slightly "T"-shaped in cross section but distinctly rounded and cordlike, almost adherent on adult specimens; thickened and rounded ribs ornamented with 2 to 4 incised spiral threads; areas between ribs smooth, unornamented; siphonale canal well developed; umbilicus narrow."
Type Specimens: HOLOTYPE: length 35 mm (incomplete), from the Sunken Meadow Member (Zone 1) of the Yorktown Formation in the PotashCorps Phosphate Mine (old Lee Creek Phosphate Mine), Aurora, North Carolina, UF 21447; PARATYPE: length 65 mm, from the same locality and formation as the holotype, in the Petuch research collection.
Type Locality: From an exposure of the Sunken Meadow Member (Zone 1) of the Yorktown Formation, in the PotashCorps Phosphate Mine (old Lee Creek Phosphate Mine), Aurora, Beaufort County, North Carolina.
Stratigraphic Range: Restricted to Zanclean Pliocene beds from North Carolina south to southern Florida, and known from the Sunken Meadow Member (Zone 1) of the Yorktown Formation and also the Jupiter Waterway Member, Murdock Formation of Florida.
Etymology: Named as a combination of "pachyus", or "thick" and "costata", or "ribbed".
Discussion: *Ecphora (Powhatan) pachycostata* represents the last-living member of its subgenus, demonstrating that the predominantly-Miocene *Powhatan* managed to survive into the Zanclean Pliocene, where it occurred together with *E. quadricostata* in the Sunken Meadow Member of the Yorktown Formation. Although "T"-shaped in cross section, the edges of the ribs of *E. (Powhatan) pachycostata* curl inward, giving the ribs a rounded, broad, cordlike appearance.

Iconography of *Powhatan* Species

Figure 5.17 ***Ecphora (Powhatan) calvertensis*** **Petuch, 1988**
A, B= typical specimen, length 43 mm. Collected in Shattuck Zone 14, 0.5 km north of Scientists' Cliffs State Park, Calvert County, Maryland; from the lower bed of the Calvert Beach Member, Calvert Formation.

Figure 5.18 *Ecphora (Powhatan) chesapeakensis* Petuch, 1988
A, B= paratype specimen, length 41 mm. Collected in Shattuck Zone 16, 0.5 km north of Scientists Cliffs State Park, Calvert County, Maryland; from the upper bed of the Calvert Beach Member, Calvert Formation.

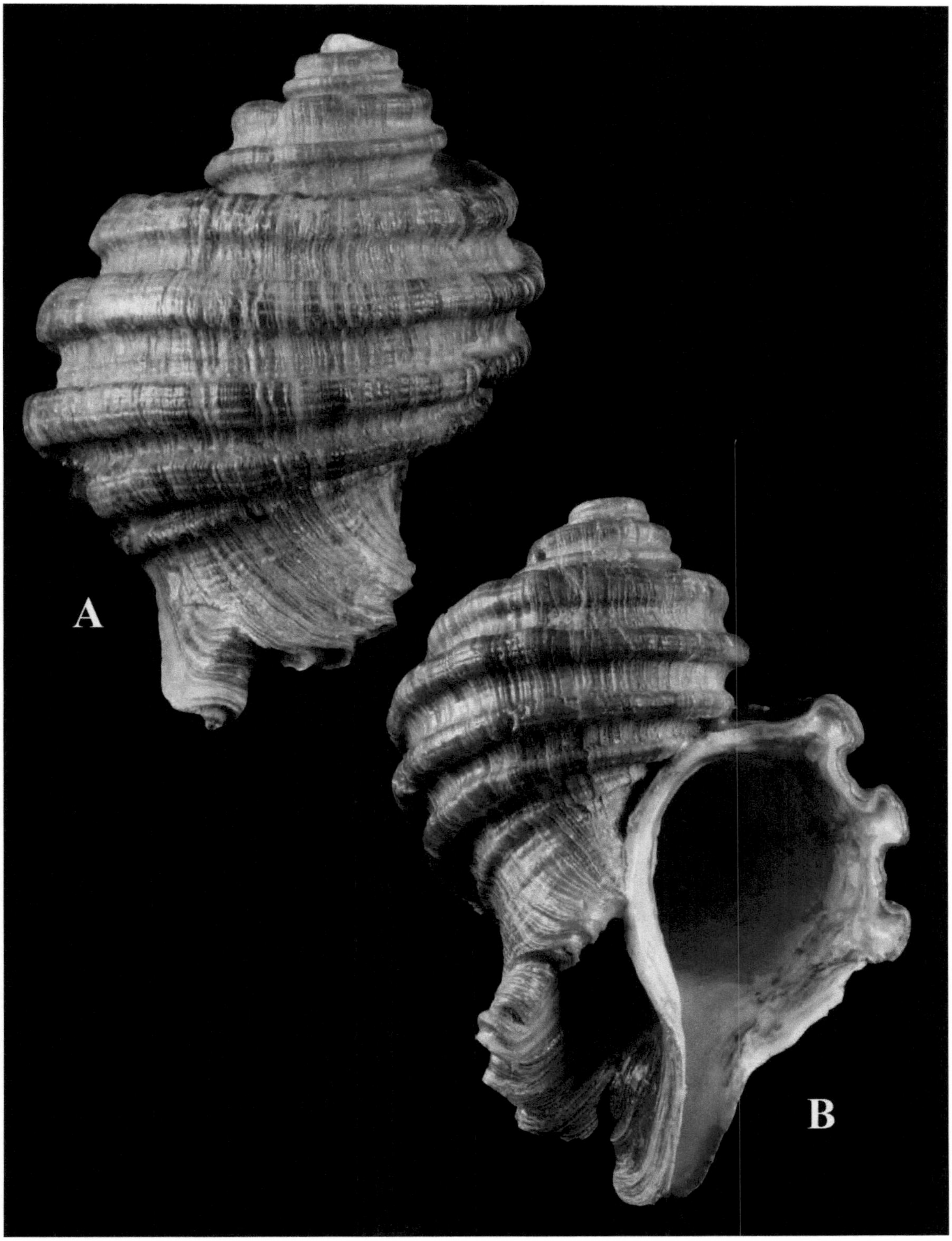

Figure 5.19 *Ecphora (Powhatan) williamsi* Ward and Gilinsky, 1988
A, B= typical specimen, length 57 mm. Collected in Shattuck Zone 19 at Drumcliff, Patuxent River, St. Mary's County, Maryland; from the Boston Cliffs Member, Choptank Formation.

Figure 5.20 ***Ecphora (Powhatan) conoyensis*** **Petuch, 2004**
A, B= paratype specimen, length 47 mm. Collected in Shattuck Zone 20 at Camp Conoy, Calvert Cliffs, Calvert County, Maryland; from the Conoy Member, St. Mary's Formation.

Figure 5.21 *Ecphora (Powhatan) asheri* Petuch, 1988
A, B= specimen showing the typical ornate sculpture, composed of five large ribs and numerous smaller ribs and cords between the main ribs, length 75 mm. Collected in Shattuck Zone 21 at the Chesapeake Ranch Club, Little Cove Point, Calvert County, Maryland; from the lower bed of the Little Cove Point Member, St. Mary's Formation. Note the three strong cords that are present between each pair of ribs. This is the most ornately sculptured of the *Powhatan* species.

Figure 5.22 ***Ecphora (Powhatan) germonae*** **Ward and Gilinski, 1988**
A, B= typical specimen, length 82 mm; **C, D=** typical specimen, length 83 mm. Both specimens were collected in Shattuck Zone 22 at the Chesapeake Ranch Club, Little Cove Point, Calvert County, Maryland; from the upper bed of the Little Cove Point Member, St. Mary's Formation.

Figure 5.23 *Ecphora (Powhatan) gardnerae* Wilson, 1987
A, B= typical specimen, length 95 mm; **C, D**= small, perfect specimen with apertural aragonite preserved, length 62 mm. Both specimens collected in Shattuck Zone 24 at Chancellor's Point, along the St. Mary's River, St. Mary's County, Maryland; from the Windmill Point Member, St. Mary's Formation.

Figure 5.24 ***Ecphora (Powhatan) gardnerae*** **Wilson, 1987.**
A, B= large specimen with a prominent healed crab break along the shoulder rib, length 120 mm. Collected in Shattuck Zone 24 at Chancellor's Point, along the St. Mary's River, St. Mary's County, Maryland; from the Windmill Point Member, St. Mary's Formation.

Figure 5.25 ***Ecphora (Powhatan) gardnerae angusticostata*** **Petuch, 1988**
A, B= paratype specimen, length 40 mm. Collected in a hard sandstone ledge cropping out suprajacent to Shattuck Zone 24, .5 km east of Chancellor Point, St. Mary's River, St. Mary's County, Maryland; from the Chancellor Point Sandstone ("Shattuck Zone 25"), St. Mary's Formation (see Petuch and Drolshagen, 2010).

Figure 5.26 ***Ecphora (Powhatan) grulkei*** **Petuch and Berschauer, new species**
A, B= holotype, length 76 mm; **C=** fragmentary specimen, length 68 mm. Both specimens were collected in the lower beds of the PotashCorps Mine, Aurora, Beaufort County, North Carolina; from the basal bed of the Claremont Manor Member, Eastover Formation, directly suprajacent to the bedding plane surface of the underlying Pungo River Formation.

Figure 5.27 *Ecphora (Powhatan) whiteoakensis* Ward and Gilinski, 1988
A, B= typical specimen, length 127 mm. Collected in the lowest bed of the cliff along the York River, south of York View, Gloucester County, Virginia; from the Claremont Member, Eastover Formation; **C, D=** typical specimen, length 82 mm. Collected in a low cliff along Urbanna Creek, Urbanna, Middlesex County, Virginia. Note the encrustations of *Septastrea* corals; from the Claremont Manor Member, Eastover Formation.

Figure 5.28 ***Ecphora (Powhatan) whiteoakensis hopei*** **Petuch, 1991**
A, B= typical specimen, length 101 mm. Collected along the Cocoplum Canal, North Port, Sarasota County, Florida; from the Cocoplum Member, Bayshore Formation.

Figure 5.29 ***Ecphora (Powhatan) pachycostata*** **Petuch, 1988**
A, B= paratype specimen, length 48 mm. Collected in the PotashCorps Mine, Aurora, Beaufort County, North Carolina; from the Sunken Meadow Member (Zone 1), Yorktown Formation; **C, D=** typical specimen, length 76 mm. Collected in a construction pit along Toledo Blade Boulevard, Murdock, Charlotte County, Florida; from the Jupiter Waterway Member, Murdock Formation.

View of an exposure of the Choptank Formation along the Calvert Cliffs near St. Leonards, Calvert County, Maryland. At this locality, *Planecphora* species such as *P. vokesi* and *P. choptankensis* are abundant and frequently exposed in the sides of the cliffs. Mrs. Joyce Spengler is shown here for scale.

CHAPTER 6. The Genus *Planecphora* Petuch, 2004

This group of thin-ribbed ecphoras converges on the shell morphology and rib structure of *Trisecphora (Matoaka)*, but differs in having four ribs instead of three. Since *Ecphora s.s.*, as *Ecphora wardi*, is already present in Shattuck Zone 10 of the Plum Point Member, the genus *Planecphora* appears to have been an evolutionary offshoot that first appeared in the Calvert Beach Member as *P. turneri* of Shattuck Zone 14. The genus later became an ecologically-dominant group in the subsequent Choptank Formation, being represented by three common species, with one in each of the three members of the Choptank Formation. By the end of Choptank deposition, marked by Shattuck Zone 19, the genus disappears from the Maryland fossil record and is not found in the subsequent St. Mary's or Eastover Formations. The genus does not appear again until the Pliocene, where it is represented by only two species, the rare Zanclean-aged *P. hertweckorum* from southern Florida and the widespread early Piacenzian *P. mansfieldi*, which is found in formations from Virginia and North Carolina south to Florida. The genus may have found a refuge in deeper offshore waters, disappearing from intertidal and coastal areas during the cold climate intervals of the late Serravallian and Tortonian Ages and only returned to shallow water areas during the warming period at the beginning of the Zanclean Age of the Pliocene.

SYSTEMATICS

Class Gastropoda Cuvier, 1797
Subclass Orthogastropoda Ponder and Lindberg, 1995
Superorder Caenogastropoda Cox, 1960
Order Sorbeoconcha Ponder and Lindberg, 1997
Infraorder Neogastropoda Wenz, 1938
Superfamily Muricoidea Rafinesque, 1815
Family Muricidae Rafinesque, 1815
Subfamily Ocenebrinae Cossmann, 1903
Tribe Ecphorini Petuch and Berschauer, new tribe
Genus *Planecphora* Petuch, 2004

Diagnosis: Ecphoras of average size, with inflated body whorls, wide shoulders, and proportionally low spires; body whorl with 4 thin, bladelike ribs that project beyond the body whorl outline; ribs always rounded on edges, never with grooves, thickenings, flanges, or flattenings. This genus was originally described as a subgenus of *Ecphora* but is considered here to be of full specific rank.

Type Species: *Planecphora choptankensis* (Petuch, 1988) (originally described as *Ecphora choptankensis*), Shattuck Zone 18, Drumcliff Member, Choptank Formation of Maryland.

Species in *Planecphora*: *Planecphora turneri* (Petuch, 1992), Calvert Beach Member (Shattuck Zone 14), Calvert Formation; *Planecphora vokesi* (Petuch, 1988), St. Leonards Member (Shattuck Zone 17), Choptank Formation; *Planecphora choptankensis* (Petuch,

1988), Drumcliff Member (Shattuck Zone 18), Choptank Formation; *Planecphora delicata* (Petuch, 1988), Boston Cliffs Member (Shattuck Zone 19), Choptank Formation; *Planecphora hertweckorum* (Petuch, 1987), Sarasota Member, Murdock Formation; *Planecphora mansfieldi* (Petuch, 1988), Rushmere Member of the Yorktown Formation, Buckingham Member of the Tamiami Formation, and Jackson Bluff Formation.

Etymology: A combination of the Latin "planum" (flat) and *Ecphora*, in reference to this group's thin, flattened ribs.

Discussion: Since there are no known *Planecphora* species in the Tortonian and Messinian formations of Maryland, Virginia, or Florida (such as the St. Mary's, Eastover, Yellow River, Red Bay, etc.), it is possible that the thin bladelike ribs of Pliocene *Planecphora* species are simply an example of convergent evolution and that these younger species do not actually belong in what could be a polyphyletic genus. Tortonian and Messinian specimens of *Planecphora* species would need to be discovered and documented in order to prove the continuity of the Maryland Miocene taxa with the Florida and North Carolina Pliocene taxa.

View of the "*Ecphora* Zone" (dark gray layer with waterfalls) of the Jackson Bluff Formation at Alum Bluff on the Apalachicola River, Florida. This is the type locality of *Planecphora mansfieldi* and many specimens were extracted from the cliff at this location.

Planecphora turneri (Petuch, 1992)
(Figure 6.1 A, B)

Description: (Reprinted from the original description, as *Ecphora (Ecphora) turneri*; Petuch, 1992: 70, figures 8, 9) "Shell cylindrical in shape, with sharply-angled shoulder; body whorl ornamented with 4 thin, narrow, bladelike ribs; edge of ribs rounded; subsutural areas flattened, producing stepped, scalariform spire; areas between ribs smooth, without spiral sculpture; siphonal canal elongated; umbilicus narrow but well developed.

Type Specimens: HOLOTYPE: length 38 mm (incomplete), in Shattuck Zone 14, 1 km north of Governor Run, Calvert Cliffs, Calvert County, Maryland, UF 21465; PARATYPE: length 40 mm, same locality and formation as the holotype, in the Petuch research collection.

Type Locality: Collected in Shattuck Zone 14, 1 km north of Governor Run, Calvert Cliffs, Calvert County, Maryland; from the Calvert Beach Member of the Calvert Formation.

Stratigraphic Range: Confined to Shattuck Zone 14 of the Calvert Beach Member, Calvert Formation, late Langhian Age of the early middle Miocene.

Etymology: Named for Joseph Turner of Baltimore, Maryland, who collected the holotype and generously donated important study material from Zones 12 and 14.

Discussion: *Planecphora turneri* is morphologically-closest to the younger and stratigraphically-higher *P. vokesi* (described next) from Shattuck Zone 17 of the Choptank Formation, but differs in being a much smaller, more cylindrical and slender shell, with lower, less projecting ribs, and a proportionally longer siphonal canal. The umbilicus of the Zone 14 species is also much narrower than that of its Zone 17 descendant.

Planecphora vokesi (Petuch, 1988)
(Figure 6.2 A, B, C)

Description: (Reprinted from the original description, as *Ecphora (Ecphora) choptankensis vokesi*; Petuch, 1988: 81, 82, figure 35) "Shell very inflated, rounded, with low spire; body whorl ornamented with four low, cordlike ribs; body and spire whorls smooth between ribs, devoid of any spiral sculpture; edge of ribs rounded, with one faint, impressed, spiral thread; umbilicus very large, flaring, well developed; siphonal canal long, broad, ornamented with few faint spiral threads."
Type Specimens: HOLOTYPE: length 86 mm, from Shattuck Zone 17 at Matoaka Cottages, St. Leonards, Calvert County; St. Leonard Member of the Calvert Formation, UF 20513; PARATYPE: length 62 mm, from the same locality and formation as the holotype, in the Petuch research collection.
Type Locality: Collected in Shattuck Zone 17 at Matoaka Cottages, St. Leonards, Calvert Cliffs, Calvert County, Maryland; St. Leonard Member of the Choptank Formation.
Stratigraphic Range: Restricted to the St. Leonard Member (Shattuck Zone 17) of the Choptank Formation, early Serravallian Age of the middle Miocene. The species is also known from Shattuck Zone 17 beds at Sandgates, St. Mary's County, Maryland, along the Patuxent River.
Etymology: Named for Dr. Harold E. Vokes of Tulane University, who authored a field guide to Maryland Miocene mollusks in 1957.
Discussion: *Planecphora vokesi* is most similar to the younger and stratigraphically-higher *P. choptankensis* (described next) but differs in being a larger shell with a broader, more inflated appearance, by having lower, less bladelike ribs, and by having a larger, much more flaring umbilicus. Because of this last feature, Martin (1904: plate 52, figure 4) incorrectly referred to this Zone 17 species as "*Ecphora quadricostata* var. *umbilicata* (Wagner)", which is now known to be a Pliocene species from the Yorktown Formation of Virginia and the Tamiami Formation of Florida.

Planecphora choptankensis (Petuch, 1988)
(Figure 6.3 A, B)

Description: (Reprinted from the original description, as *Ecphora choptankensis choptankensis*; Petuch, 1988: 82-83, figure 37) "Shell inflated, globose, thin, with 4 ribs; ribs very thin, bladelike, well-developed, extending from body whorl; siphonal canal large, broadly developed; umbilicus very large, open, flaring at anterior end; large specimens with last whorl only loosely attached, partially detached and uncoiled."
Type Specimens: HOLOTYPE: length 45 mm, width 37 mm, Matoaka Cottages, St. Leonards, Calvert County, Maryland; Shattuck Zone 18, Drumcliff Member, Choptank Formation, USNM 424248; PARATYPES: 3 specimens, 24-50 mm, same locality and stratigraphic unit as the holotype, USNM 424249; length 61 mm, same locality as the holotype, in the research collection of the senior author.
Type Locality: Shattuck Zone 18, Drumcliff Member, Choptank Formation, exposed in the cliff face at Matoaka Cottages, St. Leonard, Calvert County, Maryland.
Stratigraphic Range: Confined to Shattuck Zone 18, Drumcliff Member of the Choptank Formation, Serravallian Age, Middle Miocene.
Etymology: Named for the Choptank Formation.
Discussion: *Planecphora choptankensis* is most similar to the older *P. vokesi* from Shattuck Zone 17 but differs in being a more elongated and cylindrical shell, by having a narrower siphonal canal and a less-developed and narrower umbilicus, and by having narrower, more bladelike ribs that project farther from the body whorl. *Planecphora choptankensis* is common in the Calvert Cliffs near Matoaka Cottages, St. Leonards, Calvert County, Maryland and is frequently found as broken fragments along the beach. This species has also been collected at Sandgates, St. Mary's County, along the Patuxent River.

Planecphora delicata (Petuch, 1988)
(Figure 6.4 A, B)

Description: (Reprinted from the original description, as *Ecphora (Ecphora) choptankensis delicata*; Petuch, 1988: 83, figure 38) "Shell thin, delicate; body whorl cylindrical, with extremely well developed, bladelike ribs; edge of ribs rounded; ribs project well away from body whorl, with second rib being best developed and projecting farthest from body whorl; shoulder rib projecting upward, producing cuplike, canaliculated shoulder and spire whorls; body whorl and spire whorls sculptured with very fine spiral threads, giving shell silky texture; umbilicus narrow, poorly developed; siphonal canal narrow, elongated."
Type Specimens: HOLOTYPE: length (incomplete) 27 mm, width 30 mm, from the upper bed (Shattuck Zone 19) at Drumcliff, Jones Wharf, St. Mary's County, Maryland, along the Patuxent River, UF 20514; PARATYPES: length 52 mm, same locality and formation as the holotype, ANSP 52867; length 39 mm, same locality and formation as the holotype, in the Petuch research collection.
Type Locality: Collected in Shattuck Zone 19 at Drumcliff, Jones Wharf, along the Patuxent River, St. Mary's County, Maryland; from the Boston Cliffs Member of the Choptank Formation.
Stratigraphic Range: Confined to Shattuck Zone 19, Boston Cliffs Member of the Choptank Formation, late Serravallian Age of the late middle Miocene. The species has also been collected in Zone 19 beds at Matoaka Cottages, St. Leonards, Calvert County and at Sandgates, St. Mary's County.
Etymology: The taxon reflects the delicate, fragile appearance of this distinctive species.
Discussion: *Planecphora delicata* is most similar to the older and stratigraphically-lower *P. choptankensis* but differs in being a smaller, much more delicate and more cylindrical shell, by having better developed and more highly produced ribs, in having the shoulder rib extending upward to produce a cuplike shoulder area, and in having a much narrower and less developed umbilicus. This species represents that last-known member of *Planecphora* to have existed in Maryland, as the genus is absent from the subsequent late Miocene St. Mary's and Eastover Formations.

Planecphora hertweckorum **(Petuch, 1987)** **(Figure 6.5 A, B)**

Description: (Reprinted from the original description as *Ecphora hertweckorum*; Petuch, 1987:204, figures 14, 15) "Shell inflated, thin, delicate; body whorl with 4 very thin, greatly-projecting, blade-like spiral ribs; ribs rounded along edges, without "T"-shaped flange; shoulder sharply angled; spire elevated, scalariform; spire whorls with 2 ribs per whorl; shell smooth between ribs, without sculpturing; umbilicus wide, flaring; calcitic shell layer colored cream to yellowish-tan."
Type Specimens: HOLOTYPE: length 50 mm (fragmentary, missing siphonal canal), from Unit 11 in the APAC Pit, Sarasota, Sarasota County, Florida, Sarasota Member of the Murdock Formation, MCZ 113036; PARATYPES: 3 fragmentary specimens, lengths 35-47 mm, from the same locality and formation as the holotype, MCZ 113037; lengths 33.3 mm and 81 mm, same locality and formation as the holotype, in the research collection of the senior author.
Type Locality: In Unit 11 of the APAC Pit, Sarasota, Sarasota County, Florida; Sarasota Member of the Murdock Formation.
Stratigraphic Range: Confined to beds of late Zanclean Age, Pliocene in southern Florida, from Sarasota County, Collier County, and Okeechobee County (fragmentary specimens).
Etymology: Named for Violet ("Vi") and Charles ("Charlie") Hertweck of Venice, Florida, in recognition of their extensive fossil collecting around southern Florida, which resulted in the discovery of many new paleospecies.
Discussion: This large thin-ribbed species is the first of its genus to appear in the fossil record of Florida. The genus *Planecphora* is not present in the older, Tortonian Miocene beds of Bayshore Formation and it may have migrated into southern Florida during the early Zanclean Pliocene, possibly as a result of warming marine climates.

Planecphora mansfieldi (Petuch, 1988)
(Figures 6.6 A, B, C, D; 6.7 A, B, C, D)

Description: (Reprinted from the original description, as *Ecphora (Ecphora) mansfieldi*; Petuch, 1988: 129-130, plate A1, figures A, E) "Shell inflated, globose, with rounded body whorl; shoulder and subsutural area slightly sloping; body whorl ornamented with 4 thin, rounded, bladelike ribs; anteriormost rib thicker and wider than others; shoulder rib thinnest and most bladelike of ribs; shell smooth and unsculptured between ribs; siphonal canal wide; umbilicus well developed, flaring."
Type Specimens: HOLOTYPE: length 55 mm, from "Harvey's Creek, half a mile above abandoned mill, Leon County, Fla. (from Mansfield, 1930: plate 17, figure 7), USNM 112717; PARATYPE: lengths 85 mm and 101 mm, from Unit 10, Buckingham Member of the Tamiami Formation, in the Quality Aggregates Pit #6, Sarasota, Sarasota County, Florida, in the Petuch research collection.
Type Locality: Found along Harvey's Creek, one half mile above an abandoned mill, Leon County, Florida; in the Jackson Bluff Formation.
Stratigraphic Range: Confined to beds of early Piacenzian Age, late Pliocene, in Virginia, North Carolina, northwestern Florida, and southern Florida; in the Rushmere Member of the Yorktown Formation in Virginia and North Carolina, the lower bed of the Jackson Bluff Formation of northwestern Florida ("*Ecphora* Zone" of Mansfield, 1930), and the Buckingham Member of the Tamiami Formation of southern Florida.
Etymology: Named for Dr. Wendell C. Mansfield of the US Geological Survey, pioneer paleomalacologist and biostratigrapher of Florida and the Atlantic Coastal Plain areas.
Discussion: Mansfield (1930:70) originally referred to this distinctive, thin-ribbed species as "*Ecphora quadricostata umbilicata* (Wagner)", but noticed that the Florida species differed from true *E. quadricostata umbilicata* by lacking ribs that are "T"-shaped in cross section. This lack of flange development on the edges of the ribs demonstrates that Mansfield's illustrated specimen clearly belongs in the genus *Planecphora*. Besides being one of the most widespread ecphoras, ranging from Virginia to Florida, *P. mansfieldi* is also the last member of its genus, living well into the mid-Piacenzian Age of the Pliocene.

Iconography of *Planecphora* Species

Figure 6.1 *Planecphora turneri* (Petuch, 1992)
A, B= paratype specimen, length 40 mm. Collected in Shattuck Zone 14, 1 km north of Governor Run, Calvert Cliffs, Calvert County, Maryland; from the Calvert Beach Member, Calvert Formation.

Figure 6.2 *Planecphora vokesi* (Petuch, 1988)
A, B= paratype specimen, length 62 mm. Collected in Shattuck Zone 17 at Matoaka Cottages, St. Leonards, Calvert Cliffs, Calvert County, Maryland; from the St. Leonard Member, Choptank Formation; **C=** large fragmentary specimen, length 69 mm. Collected in Shattuck Zone 17 at Sandgates, along the Patuxent River, St. Mary's County, Maryland.

Figure 6.3 *Planecphora choptankensis* (Petuch, 1988)
A, B= paratype specimen, length 62 mm. Collected in Shattuck Zone 18, between Matoaka Cottages and Governor Run, St. Leonard, Calvert Cliffs, Calvert County, Maryland; from the Drumcliff Member, Choptank Formation.

Figure 6.4 *Planecphora delicata* (Petuch, 1988)
A, B= paratype specimen, length 39 mm. Collected in Shattuck Zone 19 at Drumcliff, along the Patuxent River, St. Mary's County, Maryland; from the Boston Cliffs Member, Choptank Formation.

Figure 6.5 *Planecphora hertweckorum* (Petuch, 1987)
A, B= fragmentary paratype specimen, length 81 mm. Collected in Unit 11 in Quality Aggregates Pit #6, Sarasota, Sarasota County, Florida; from the Sarasota Member of the Murdock Formation.

Figure 6.6 *Planecphora mansfieldi* (Petuch, 1988). Specimens from northwestern Florida and North Carolina. A, B= typical specimen, length 47 mm. Collected in the middle bed at Alum Bluff, along the Apalachicola River, Liberty County, Florida; from the lower bed of the Jackson Bluff Formation; similar to Mansfield's illustrated specimen (1930: plate 17, figure 7); **C, D**= typical specimen, length 66 mm. Collected in the middle beds of the PotashCorps Mine, Aurora, Beaufort County, North Carolina; from the Rushmere Member, Yorktown Formation.

Figure 6.7 *Planecphora mansfieldi* (Petuch, 1988). Specimens from southern Florida.
A, B= paratype specimen, length 85 mm; **C, D**= typical specimen, length 80 mm. Both specimens collected in Unit 10 in Quality Aggregates Pit #6, Sarasota, Sarasota County, Florida; from the Buckingham Member, Tamiami Formation. The ribs on specimens of this species that are found in southern Florida are thinner and more delicate than those on specimens from northern Florida and may represent an undescribed regional subspecies.

Globecphora parvicostata specimen, 67 mm, from the Goose Creek Formation.

CHAPTER 7. The Genus *Globecphora* Petuch, 1994

Although similar to some *Ecphora* s.s. and *E. (Rhombecphora)* species, *Globecphora* species differ in having much more inflated shells with reduced rib development. Although some *Ecphora (Powhatan)* species may attain lengths and sizes comparable to those of *Globecphora*, they still retain their strong ribs and never exhibit the great degree of body whorl inflation seen in *Globecphora*.

SYSTEMATICS

Class Gastropoda Cuvier, 1797
Subclass Orthogastropoda Ponder and Lindberg, 1995
Superorder Caenogastropoda Cox, 1960
Order Sorbeoconcha Ponder and Lindberg, 1997
Infraorder Neogastropoda Wenz, 1938
Superfamily Muricoidea Rafinesque, 1815
Family Muricidae Rafinesque, 1815
Subfamily Ocenebrinae Cossmann, 1903
Tribe Ecphorini Petuch and Berschauer, new tribe
Genus *Globecphora* Petuch, 1994

Diagnosis: Shells very large for its Tribe, among the largest species; very inflated, globose, rounded, with wide, flaring apertures; many species with 5 ribs, with the anteriormost rib most often being smaller or almost obsolete; ribs proportionally small, narrow, generally low flattened; some species with ribs being virtually obsolete, no more than undulations on surface; umbilicus open, extremely flared, very wide; early whorls and juvenile often similar to *Ecphora* s.s. species in shape and rib structure; last whorl of mature specimens widely expanded, characteristically inflated and rotund.
Type Species: "*Ecphora*" *floridana* Petuch, 1988, Port Charlotte Member, Bayshore Formation, late Messinian Miocene of southern Florida.
Species in *Globecphora*: *Globecphora floridana* (Petuch, 1988), Port Charlotte Member, Bayshore Formation; *G. floridana streami* Petuch, 1994, uppermost beds of the Port Charlotte Member, Bayshore Formation; *G. parvicostata* (Pilsbry, 1911), Sunken Meadow Member of the Yorktown Formation, Goose Creek Formation, and Buckingham Member of the Tamiami Formation (*G. xenos* Petuch is a synonym); *G. mardieae* Petuch and Berschauer, n. sp., Morgarts Beach Member of the Yorktown Formation.

Globecphora floridana (Petuch, 1988)
(Figure 7.1 A, B)

Description: (Reprinted from the original description, as *Ecphora (Ecphora) floridana*; Petuch, 1988: 67) "Shell extremely inflated, globose, with rounded shoulder and sloping subsutural area; body whorl ornamented with 5 low, cordlike ribs; ribs adherent, rounded in cross section; ribs more elevated and pronounced on early whorls, becoming lower and more adherent on later whorls; ribs ornamented with two deeply impressed threads; areas between ribs heavily sculptured with numerous fine spiral threads; umbilicus wide and flaring."

Type Specimens: HOLOTYPE: length 92 mm, from Unit 11 in the APAC Pit, Sarasota, Sarasota County, Florida, MCZ 29452; PARATYPES: length 30 mm, from the same locality and formation as the holotype, MCZ 29453; length 110 mm, from a construction site at El Jobean, Charlotte County, Florida, Port Charlotte Member of the Bayshore Formation, in the research collection of the senior author.

Type Locality: In a deep drainage ditch, below Unit 11 ("Unit 12") in the APAC Pit, Richardson Road, Sarasota, Sarasota County, Florida, in the Port Charlotte Member of the Bayshore Formation.

Stratigraphic Range: Confined to the Port Charlotte Member of the Bayshore Formation of southern Florida, late Messinian Age of the late Miocene. Known only from Sarasota and the Port Charlotte areas of southwestern Florida.

Etymology: Named for the State of Florida.

Discussion: This highly inflated and globose species is the largest ecphora to have lived in Florida during the Miocene, with some individuals reaching the size of a small cantaloupe. It is also the oldest-known *Globecphora* species and the first of its genus to appear in the fossil record.

Globecphora floridana streami (Petuch, 1994)
(Figure 7.2 A, B)

Description: (Reprinted from the original description; Petuch, 1994: 290, plate 50, figure B) "Shell of average size for genus, with general shell characters conforming to generic description; ribs well developed for genus, narrow; intercostal areas heavily sculptured with numerous strong spiral cords, with central 2 or 3 cords being largest; anterior fifth rib well-developed, almost as large as other 4 ribs."
Type Specimens: HOLOTYPE: length 76 mm, from the deepest exposures in the APAC Pit, Sarasota, Sarasota County, Florida, uppermost bed of the Port Charlotte Member, Bayshore Formation, UF 66271.
Type Locality: Collected from the bottom of a deep drainage channel, 3 m below the base of Unit 11 in the APAC Pit, Sarasota, Sarasota County, Florida; from the uppermost bed of the Port Charlotte Member, Bayshore Formation (considered to be "Unit 12" in the APAC Pit).
Stratigraphic Range: Confined to beds of latest Messinian Age, late Miocene and known only from the upper beds of the Port Charlotte Member of the Bayshore Formation, from Sarasota, Sarasota County, Florida.
Etymology: Named for Dale Stream of Sarasota, Florida, who collected the holotype and who kindly donated large suites of ecphoras for study.
Discussion: *Globecphora floridana streami* is most similar to the older and stratigraphically-lower *G. floridana floridana* but differs in being a smaller, more cylindrical and higher-spired shell, in having narrower and more projecting ribs, and most importantly, in having numerous prominent and conspicuous secondary cords between the four main cords on the body whorl, shoulder, and spire whorls. Because of these numerous strong secondary cords, this late Miocene shell is the most ornately-sculptured ecphora found in the Florida fossil record.

Globecphora parvicostata (Pilsbry, 1911)
(Figure 7.3 A, B, C, D)

Description: (Reprinted from the original description, originally named as *Ecphora parvicostata*; Pilsbry, 1911:438-439) "The shell is shaped like the more globose forms of *Ecphora quadricostata* (Say): the umbilicus moderate, rapidly contracting inward, not so ample as in *E. q. umbilicata* (Wagner). The last whorl has four very low, narrow spiral ribs: the surface above the upper rib slopes upward to the suture, immediately below which it is a little swollen. The spiral ribs are stronger on the spire, where two are exposed, the suture revolving on the third. Between the ribs there are irregular growth-wrinkles, but no spiral striae." (*Ecphora (Trisecphora) xenos* Petuch, 1989, based on a fragment, is now known to be a synonym).
Type Specimens: HOLOTYPE: length 82 mm, width 70 mm, ANSP 1514.
Type Locality: Pilsbry had no definite type locality, merely stating "Chesapeake Miocene". The species is now known to be Pliocene in age and probably was originally collected in the Yorktown Formation of Virginia.
Stratigraphic Range: Based upon specimens subsequently collected in Virginia, North Carolina, and South Carolina, the species is now known to be early Piacenzian Pliocene in age. In Virginia and North Carolina, this species is confined to the Rushmere Member of the Yorktown Formation (lower Zone 2 Yorktown); in South Carolina it is common in the lower beds of the Goose Creek Formation.
Etymology: A combination of the Latin "parvus", meaning "small or unimportant" and the Latin "costatus", ribbed; the Small-Ribbed Ecphora.
Discussion: This distinctive globose and inflated species has a wide range in early Pliocene beds along the entire southeastern United States, being found from Virginia, North and South Carolina, and southern Florida. It is most similar to the older Messinian Miocene *G. floridana*, but differs in being a consistently smaller species with a wider and more sloping shoulder and in having thinner ribs that are more adherent to the body whorl. The Florida Miocene species consistently has five distinct ribs on the body whorl, while its Pliocene descendant only ever has four. The senior author named "*Ecphora (Trisecphora)" xenos* (Petuch, 1989: 134, plate A3, figure C) from a broken fragment, conjecturing that it represented a new three-ribbed North Carolina Miocene species. More complete specimens, collected subsequently to the fragmentary holotype (such as the specimen shown here on Figure 7.3 A, B), have shown that *xenos* is actually a four-ribbed Pliocene species and that it is a synonym of Pilsbry's *parvicostata*.

Globecphora mardieae Petuch and Berschauer, new species
(Figure 7.4 A, B, C, D)

Description: Shell small for genus, extremely inflated, globose, with rounded shoulder and sloping subsutural area; body whorl ornamented with 4 low, cordlike ribs; fifth smaller and less-developed rib around base of siphonal canal; ribs adherent, rounded in cross section; ribs more elevated and pronounced on early whorls, becoming lower and more adherent on later whorls; ribs ornamented with two deeply impressed threads; 4 main ribs unequal in size, thickness and width gradational, ranging from narrow, small shoulder rib and grading to very wide anterior rib, which is over twice as thick as shoulder rib; areas between ribs smooth, only faintly ornamented with few fine spiral threads; aperture wide and round in shape; siphonal canal short, stumpy, and broad, ornamented with 6-8 small, low cords; umbilicus narrow and poorly developed.
Type Specimens: HOLOTYPE: length 41 mm, from the middle Yorktown beds in the PotashCorps Phosphate Mine (old Lee Creek Mine), Aurora, North Carolina, PRI 104779; PARATYPE: length 68 mm, same locality and formation as the holotype, in the Petuch research collection.
Type Locality: Collected in the upper section of the middle beds in the PotashCorps Mine, Aurora, Beaufort County, North Carolina; from the Morgarts Beach Member, Yorktown Formation.
Stratigraphic Range: Confined to the Morgarts Beach Member of the Yorktown Formation (middle Zone 2), early Piacenzian Age, Pliocene. To date, known only from the type locality.
Etymology: Named for Mardie Drolshagen of Stecoah Township, North Carolina, in recognition of her many contributions to the molluscan paleontology of the Chesapeake Miocene.
Discussion: *Globecphora mardieae* is unique among the members of *Globecphora* in that it has noticeably unequal rib sizes, with the shoulder rib being only half the width of the anterior rib and with the other two ribs being gradational between them. This inequicostate pattern is especially obvious when examining the outer edge of the aperture, where the anteriormost rib (bottom rib) is at least twice as wide as the thin and narrow shoulder rib. *Globecphora mardieae* was the last-living member of its genus, as *Globecphora* disappears from the fossil record after Morgarts Beach time.

Iconography of *Globecphora* Species

Figure 7.1 ***Globecphora floridana*** **(Petuch, 1988)**
A, B= paratype specimen, length 110 mm. Collected at a housing development construction site in El Jobean, Charlotte County, Florida; from the Port Charlotte Member, Bayshore Formation.

Figure 7.2 ***Globecphora floridana streami*** **(Petuch, 1994)**
A, B= typical specimen, showing the characteristic multiple heavy cords between the main ribs, length 112 mm. Collected from the bottom of a deep drainage channel, 3 m below the base of Unit 11 ("Unit 12") in the APAC Pit, Sarasota, Sarasota County, Florida; from the uppermost bed of the Port Charlotte Member, Bayshore Formation.

Figure 7.3 *Globecphora parvicostata* (Pilsbry, 1911)
A, B= typical specimen, length 80 mm. Collected in the middle beds of the PotashCorps Mine, Aurora, Beaufort County, North Carolina; from the Rushmere Member, Yorktown Formation. **C, D=** typical specimen, 57 mm. Collected in the Martin Marietta Berkeley Quarry, east of Eutawville, Orangeburg County, South Carolina; from the lower beds of the Goose Creek Formation.

Figure 7.4 ***Globecphora mardieae*** **Petuch and Berschauer, new species**
A, B= holotype, length 41 mm; **C, D=** paratype with well-developed fifth rib, length 68 mm (missing spire). Both specimens were collected in the upper section of the middle beds in the PotashCorps Mine (old Lee Creek Phosphate Mine), Aurora, Beaufort County, North Carolina; from the Morgarts Beach Member, Yorktown Formation.

Globecphora parvicostata (Pilsbry, 1911) wide-ribbed variant, length 85 mm, from the Rushmere Member (Zone 2) of the Yorktown Formation in the PotashCorps Phosphate Mine (old Lee Creek Mine), Aurora, North Carolina.

CHAPTER 8. The Genus *Latecphora* Petuch, 1989

The genus *Latecphora* represents the ultimate in rib development for the entire Tribe Ecphorini. Species such as *L. bradleyae* and *L. hertwecki* have the widest ribs of any known ecphora. In some specimens of *L. bradleyae*, the ribs actually touch (as seen here on Figure 8.2C, D), in effect coalescing to produce a double shell. This rib fusion may have evolved as a response to intense crab predation in the tropical seas of the Pliocene southern Florida and the Everglades Pseudoatoll coral reef complexes (see Petuch and Berschauer, 2021 for details of the coral reef structure and fauna). By having coalescing rib edges, the shell is, in effect, doubled in thickness without doubling the weight. This would have made it more difficult for peeling crabs, such as *Menippe* and *Calappa*, to break the shells of the advanced *Latecphora* species. Based on its stratigraphic position at the top of the Pinecrest Member of the Tamiami Formation, *Latecphora hertwecki* was the last-living ecphora and the entire Tribe became extinct after Pinecrest time.

SYSTEMATICS

Class Gastropoda Cuvier, 1797
Subclass Orthogastropoda Ponder and Lindberg, 1995
Superorder Caenogastropoda Cox, 1960
Order Sorbeoconcha Ponder and Lindberg, 1997
Infraorder Neogastropoda Wenz, 1938
Superfamily Muricoidea Rafinesque, 1815
Family Muricidae Rafinesque, 1815
Subfamily Ocenebrinae Cossmann, 1903
Tribe Ecphorini Petuch and Berschauer, new tribe
Genus *Latecphora* Petuch, 1988 (originally described as a subgenus of *Ecphora*)

Diagnosis: (from Petuch, 1988: 103-104) "Shells similar to *Ecphora*, being globose, inflated, with wide apertures and elevated spires; shoulders sloping, especially in advanced forms, giving shells distinct turnip shape; body whorls ornamented with four extremely-wide ribs that are extensively "T"-shaped in cross section; edges of ribs in advanced forms almost coalesce to produce a second shell layer; siphonal canals short and stubby in proportion to shell sizes; umbilici generally extremely open and flaring, especially in advanced forms."
Type Species: *Latecphora bradleyae* (Petuch, 1987), Pinecrest Member, Tamiami Formation of Florida and Duplin Formation of North and South Carolina, Piacenzian Age, Pliocene.
Species in *Latecphora*: *Latecphora violetae* (Petuch, 1988), Buckingham Member, Tamiami Formation and Goose Creek Formation; *Latecphora bradleyae* (Petuch, 1987), Pinecrest Member, Tamiami Formation and Duplin Formation; *Latecphora hertwecki* (Petuch, 1988), uppermost bed of the Pinecrest Member, Tamiami Formation.
Etymology: Named as a combination of "latus", Latin for "wide or broad" and ecphora, in reference to the characteristic ultra-wide ribs seen on members of the genus.

Latecphora violetae (Petuch, 1988)
(Figure 8.1 A, B, C)

Description: (Reprinted from the original description, as *Ecphora violetae*; Petuch, 1988b: 35-36) "Shell inflated, globose, with 4 wide spiral ribs; ribs "T"-shaped in cross section; shoulder angle sharply defined; spire elevated, protracted, with turriculate appearance; spire whorls with only 2 ribs; ribs of spire and body whorl widely-separated, with distance between ribs being equal to width of single rib; shoulder rib elevated above plane of shoulder, with posterior edge of rib extending upward in bladelike fashion; area between suture and shoulder rib flat, wide, equal to 2 ribs at edge of aperture; umbilicus well-developed, slightly flaring."

Type Specimens: HOLOTYPE: length 26 mm, width 22 mm, from Unit 10 in the APAC quarry, Sarasota, Sarasota County, Florida, Buckingham Member of the Tamiami Formation, USNM 427957; PARATYPE: length 98 mm, from the same locality and formation as the holotype, in the Petuch research collection.

Type Locality: The old APAC quarry (Ashland Petroleum and Asphalt Corporation), off Richardson Road, Sarasota, Sarasota County, Florida (now flooded as a recreational lake in a county park).

Stratigraphic Range: Confined to the early Piacenzian Age of the Pliocene, and found in the Buckingham Member (Unit 10) of the Tamiami Formation of Florida and in the lower beds of the Goose Creek Formation at Santee, South Carolina.

Etymology: Named for Violet Hertweck, loving wife of Charles Hertweck, of Venice, Florida, who collected the holotype in the APAC pit.

Discussion: This is the oldest species of the genus and it makes its first appearance in the Florida fossil record in the basal beds of the Buckingham Member of the Tamiami Formation. Rare examples of this distinctive species are also known from the Goose Creek Formation in the Martin Marietta Berkely quarry in South Carolina. Although similar in body shape and size to the younger members of the genus, *Latecphora violetae* differs in having much narrower ribs than the subsequent taxa and in having a much wider and more open subsutural area between the shoulder rib and the spire whorl sutures.

Latecphora bradleyae (Petuch, 1987)
(Figure 8.2 A, B, C, D)

Description: (Reprinted from the original description; originally described as *Ecphora bradleyae*; Petuch, 1987: 204, figures 1-6) "Shell turbinate in form, thickened, with 4 wide, flattened spiral ribs on body whorl; ribs strongly "T"-shaped in cross section; some specimens with ribs almost touching along edges; ribs ornamented with 4-8 fine, incised grooves; shell smooth and unornamented between and beneath grooves; shoulder rounded; spire elevated, with sloping outline; columella adherent; umbilicus open, flaring; siphonal canal well developed; aperture moderate in size for genus round in shape; spire whorls with 2 ribs per whorl; suture and edge of shoulder rib separated by wide space, giving spire tabulate appearance; calcitic outer shell layer colored cream-tan or reddish gray.
Type Specimens: HOLOTYPE: length 66 mm, APAC Pit (old Macasphalt Newburn Pit Mine), Sarasota, Sarasota County, Florida, USNM 358548; PARATYPES: 3 specimens, lengths 33 mm, 76 mm, and 95 mm, MCZ 29447, MCZ 29449, and MCZ 29450, same locality as the holotype.
Type Locality: In Unit 7 in the APAC Pit, Sarasota, Sarasota County, Florida, Pinecrest Member of the Tamiami Formation.
Stratigraphic Range: Confined to beds of middle Piacenzian Age, Pliocene of the southeastern United States, from South Carolina (Duplin Formation) to southern Florida (Pinecrest Member, Tamiami Formation).
Etymology: Named for Evelyn Bradley of Bradenton, Florida, who collected the holotype at Sarasota.
Discussion: The flanges on the edges of the "T"-shaped ribs of *Latecphora bradleyae* almost coalesce, in effect producing "I"-beams and doubling the shell thickness. This appears to have been an adaptation against crab predation; doubling the effective shell thickness without doubling the shell weight and would have made it more difficult for shell-peeling crabs to break back the outer lip.

Latecphora hertwecki (Petuch, 1988)
(Figure 8.3 A, B)

Description: (Reprinted from the original description, as *Ecphora (Latecphora) hertwecki*; Petuch, 1988: 105-106, figure 48) " Shell small for genus, turbinate, fusiform, with high spire; shoulder and spire whorls very sloping, producing rounded outline; body whorl ornamented with four extremely wide, "T"-shaped ribs; ribs very close together, almost touching and sometimes overlapping; subsutural area very narrow, equal to width of spaces between ribs on body whorl; subsutural area partially covered by sloping shoulder rib; siphonal canal short, stubby; umbilicus wide and flaring; ribs ornamented with six to ten fine, impressed grooves; areas between ribs smooth, without spiral sculpture."
Type Specimens: HOLOTYPE: length 33 mm, from the uppermost bed of Unit 7, Pinecrest Member, Tamiami Formation in the APAC Pit, Sarasota, Sarasota County, Florida, MCZ 29447.
Type Locality: From the bedding plane surface of the uppermost bed of Unit 7, Pinecrest Member, Tamiami Formation, in the APAC Pit, Sarasota, Sarasota County, Florida.
Stratigraphic Range: Confined to the uppermost beds of the Pinecrest Member of the Tamiami Formation, late Piacenzian Age, late Pliocene. Known only from Sarasota.
Etymology: Named for Charles ("Charlie") Hertweck of Venice, Florida who generously shared his knowledge and specimens with hundreds of amateur and professional paleontologists for decades.
Discussion: *Latecphora hertwecki* differs from the older and stratigraphically-lower *L. bradleyae* in being a much smaller and more turbinate shell with a distinctly more rounded shoulder and subsutural area. In *L. bradleyae,* the subsutural channel is proportionally wide and open, with a wide gap between the edge of the shoulder rib and the spire whorls. In the younger *L. hertwecki*, the subsutural channel is almost closed off and is only as wide as the gap between the ribs on the body whorl. This dwarf *Latecphora* species is the last-living ecphora and the entire Tribe disappears from the fossil record after the deposition of the Pinecrest Member. This once-widespread and species-rich group of muricids survived only as a single dwarf species that was restricted to the lagoons along late Pliocene western Florida.

Iconography of *Latecphora* Species

Figure 8.1 ***Latecphora violetae*** **(Petuch, 1988)**
A, B= typical specimen, length 98 mm. Collected in Unit 10 of the APAC Pit, Sarasota, Sarasota County, Florida; from the Buckingham Member, Tamiami Formation; **C**= specimen in a phosphatic limestone matrix, length 84 mm. Collected in the lowest beds of the Martin Marietta Berkeley Quarry, east of Eutawville, Orangeburg County, South Carolina; from the lower beds of the Goose Creek Formation.

Figure 8.2 ***Latecphora bradleyae*** **(Petuch, 1987)**
A, B= typical specimen, length 68 mm; **C, D=** specimen with exceptionally wide ribs and barnacles growing on the spire whorls, length 93 mm. Both specimens were collected in Unit 7 of the APAC Pit, Sarasota, Sarasota County, Florida. Also found in the Duplin Formation of South Carolina.

Figure 8.3 ***Latecphora hertwecki*** **(Petuch, 1988)**
A, B= typical specimen, length 30 mm, Collected at the top of Unit 7 of the APAC Pit, Sarasota, Sarasota County, Florida; from the uppermost bed of the Pinecrest Member, Tamiami Formation. This dwarf species is the last-living member of the genus *Latecphora* and is also the last-living ecphora, as the entire Tribe Ecphorini becomes extinct after Pinecrest time.

View of the Calvert Cliffs as seen from Matoaka Cottages, St. Leonards, Calvert County, Maryland, showing three intrepid ecphora collectors, on the beach at the base of the cliff, heading home at dusk.

References

Abbott, D.H., A.A. Nunes, I.S. Leung, and J.T. Hagstrum. 2003. The Ewing Impact Structure Progress Report. Abstract #P52A-0472, Fall Meeting 2003. American Geophysical Union.

Bouchet, P., J-P. Rocroi, B. Hausdorf, A. Kaim, Y. Kano, A. Nutzel, P. Parkhaev, M. Schrodl, and E. Strong. 2017. Revised Classification, Nomenclator, and Typification of Gastropod and Monoplacophoran Families. *Malacologia* 61 (1-2):1-526 (2017).

Campbell, L.D. 1993. *Pliocene Molluscs from the Yorktown and Chowan River Formations of Virginia.* Virginia Division of Mineral Resources, Publication 127, Charlottesville, VA. 259 pp.

Carter, J.G., T.J. Rossbach, K.J. Robertson, and L.W. Ward. 1994. Morphological and Microstructural Evidence for the Origin and Early Evolution of *Ecphora* (Mollusca: Gastropoda). *Journal of Paleontology* 68 (4, July 1994): 905-907.

Conrad, T.A. 1843. Descriptions of a new genus, and twenty-nine new Miocene, and one Eocene fossil shells of the United States. *Proceedings of the Academy of Natural Sciences of Philadelphia* 1(3): 305-311.

Dall, W.H. 1915. A Monograph of the Molluscan Fauna of the *Orthaulax pugnax* Zone of the Oligocene of Tampa, Florida. *U.S. National Museum Bulletin* 90: 1-173.

Gorter, J.D., and S.W. Bayford. 2000. Possible impact origin for the middle Miocene (Serravallian) Puffin Structure, Ashmore Platform, Australia. *Australian Journal of Earth Science* 47:707-714.

Gurov, E.P. and H.P. Gurova. 1993. The Kara-Kul Depression in the Pamirs; a reliable impact structure. *Geologi Zhurnal* 6: 53-64 (Russian Academy of Sciences).

Leung, I.S., R. Winston, and D.H. Abbott. 2004. Silicon carbide from the Canyon Diablo meteorite and the Ewing Impact Structure. Abstract #P33C-10, Spring Meeting 2004, American Geophysical Union.

Mansfield, W.C. 1939. Notes on the Upper Tertiary and Pleistocene Mollusks of Peninsular Florida. State of Florida Department of Conservation, *Geological Bulletin* No. 15. 75 pp.

Martin, G.C. 1904. Systematic Paleontology, *Miocene*, Gastropoda. *In* Clark, W.B., G.B. Shattuck, and W.H. Dall (editors), *The Miocene Deposits of Maryland.* Maryland Geological Survey, *Miocene*. pp. 131-270.

Oleinik, A.E., L. Marinkovich, K.B. Barinov, and P.K. Swart. 2008. Magnitude of middle Miocene warming in north Pacific high latitudes: stable isotope evidence from *Kaneharaia* (Bivalvia: Dosiniinae). *Bulletin of the Geological Survey of Japan* 59: 339-353.

Olsson, A.A. and R.E. Petit. 1964. Some Neogene Mollusca from Florida and the Carolinas. *Bulletins of American Paleontology* 47(217): 556-561.

Petuch, E.J. 1982. Geographical Heterochrony: Contemporaneous Co-existence of Neogene and Recent molluscan faunas in the Americas. *Palaeogeography, Palaeoclimatology, and Palaeoecology* 37: 277-312.

Petuch, E.J. 1986. The Pliocene Reefs of Miami: Their Geomorphological Significance in the Evolution of the Atlantic Coastal Ridge, Southeastern Florida, USA. *Journal of Coastal Research* 2(4): 391-408.

Petuch, E.J. 1987. A New *Ecphora* Fauna from Southern Florida. *The Nautilus* 101(4): 200-206.

Petuch, E.J. 1988a. *Neogene History of Tropical American Mollusks: biogeography and evolutionary patterns of tropical western Atlantic Mollusca.* The Coastal Education and Research Foundation, Charlottesville, Virginia. 217 pp.

Petuch, E.J. 1988b. New Species of *Ecphora* and Ecphorine Thaidids from the Miocene of Chesapeake Bay, Maryland, USA. *Bulletin of Paleomalacology* 1 (1): 1-16. (copies available in the Dall Library, Department of Malacology, Smithsonian Institution).

Petuch, E.J. 1988c. New Species of *Trisecphora* from the Miocene of Maryland, USA. *Bulletin of Paleomalacology* 1(2): 22-36. (copies available in the W.H. Dall Library, Division of Mollusks, Smithsonian Institution).

Petuch, E.J. 1988d. A New *Ecphora* from the Buckingham Formation at Sarasota, Florida, USA. *Bulletin of Paleomalacology* 1(2): 36-39. (copies available in the W.H. Dall Library, Division of Mollusks, Smithsonian Institution).

Petuch, E.J. 1988e. *Field Guide to the Ecphoras*. Coastal Education and Research Foundation, Charlottesville, Virginia. 140 pp.

Petuch, E.J. 1993. Patterns of Diversity and Extinction in Transmarian Muricacean, Buccinacean, and Conacean Gastropods. *The Nautilus* 106(4): 155-173.

Petuch, E.J. 1995. Molluscan Diversity in the Late Neogene of Florida: evidence for a two-staged mass extinction. *Science* 270; 275-277.

Petuch, E.J. 2004. *Cenozoic Seas: The View from Eastern North America.* CRC Press, London, New York, Boca Raton. 308 pp.

Petuch, E.J. 2012. Miocene Asteroid Impacts: Proposed Effects on the Biogeography and Extinction Patterns in Eastern North American Gastropods. In *Earth and Life: Global Biodiversity, Extinction, and Biogeographic Perturbations Through Time.* John A. Talent, Editor. Springer Science Media, Dordrecht, Heidelberg, London, and New York. pp. 967-981.

Petuch, E.J. and D.P. Berschauer. 2021. *Ancient Seas of Southern Florida: The Geology and Paleontology of the Everglades Region.* CRC Press, London, New York, Boca Raton. 266 pp.

Petuch, E.J. and M. Drolshagen. 2010. *Molluscan Paleontology of the Chesapeake Miocene.* CRC Press, London, New York, Boca Raton. 160 pp.

Petuch, E.J. and C.E. Roberts. 2007. *The Geology of the Everglades and Adjacent Areas.* CRC Press, London, New York, Boca Raton. 212 pp.

Pilsbry, H.A. 1911. A New *Ecphora* from the Chesapeake Miocene. *Proceedings of the Academy of Natural Sciences of Philadelphia* LXIII: 438-439.

Ray, C.E. 1987. Foreword to the *Geology and Paleontology of the Lee Creek Mine, North Carolina*, II. Smithsonian Contributions to Paleobiology 61: 1-8.

Say, T. 1824. An account of some of the fossil shells of the Academy of Natural Sciences of Philadelphia. *Proceedings of the Academy of Natural Sciences of Philadelphia* 4(1): 124-155.

Shattuck, G.B. 1904. Geological and Paleontological Relations, with a Review of Earlier Investigations. *In* Clark, W.B., G.B. Shattuck, and W.H. Dall (editors), *The Miocene Deposits of Maryland.* Maryland Geological Survey, *Miocene*. pp. xxxiii-clv.

Shoemaker, E.M. and E.C.T. Chao. 1961. New Evidence for the impact origin of the Ries Basin, Bavaria, Germany. *Journal of Geophysical Research* 66: 3371-3378.

Tuomey, M. and F.S. Holmes. 1856. *Pleiocene Fossils of South-Carolina.* Russell and Jones Printers, Charleston, South Carolina. 152 pp.

Wagner, W. 1839. Description of five new Fossils, of the older Pliocene formation of Maryland and North America. *Journal of the Academy of Natural Sciences of Philadelphia* (8, first series): 51-53. (Also read before the assembled Academy, during the January, 1838 meeting.)

Ward, L.W. 1985. *Stratigraphy and Characteristic Mollusks of the Pamunkey Group (Lower Tertiary) and the Old Church Formation of the Chesapeake Group, Virginia Coastal Plain.* U.S. Geological Survey Professional paper 1346, 78 pp, figures 1-6.

Ward, L.W. 1998. Mollusks from the Lower Miocene Pollack Farm site, Kent County, Delaware: A Preliminary Analysis. In *Geology and Paleontology of the Lower Miocene Pollack Farm Fossil Site, Delaware*, R.N. Benson, Editor. Special Publication No. 21, Delaware Geological Survey, University of Delaware, Newark, Delaware. pp. 98-131, 19 plates.

Ward, L.W. and N. Gilinsky. 1988. *Ecphora* (Gastropoda: Muricidae) from the Chesapeake Group of Maryland and Virginia. *Notulae Naturae* 469 (15 March): 1-21.

Whitfield, R.P. 1894. Mollusca and Crustacea of the Miocene Formations of New Jersey. *U.S. Geological Survey Monograph* 24. 195 pp.

Wilson, D. 1987. New Species of *Ecphora*, including the subgenus *Stenomphalus*, in the Pungo River Formation. *In* C.E. Ray, editor. Geology and Paleontology of the Lee Creek Mine, North Carolina II. *Smithsonian Contributions to Paleobiology* 61: 21-29.

Trisecphora (Matoaka) bartoni from Shattuck Zone 16, Calvert Beach Member, Calvert Formation.

Systematic Index (arranged by species)

S

T

V

W

Chesathais lindae Petuch, 1988, length 59 mm. Collected north of Matoaka Cottages, St. Leonards, Calvert County, Maryland

Index of Ecphora-Bearing Geological units

J

K

L

M

P

R

S

T

W

Y

About the Authors

Edward J. Petuch was born in Bethesda, Maryland, in 1949. Raised in a Navy family, he spent many of his childhood years collecting living and fossil shells in such varied localities as Chesapeake Bay, California, Puerto Rico, and Wisconsin. His early interests in malacology and oceanography eventually led to BA and MS degrees in zoology from the University of Wisconsin-Milwaukee. During his MS thesis research, Petuch concentrated on the molluscan biogeography of West Africa, traveling extensively in the Canary Islands, Western Sahara, Senegal, Gambia, Sierra Leone, Ivory Coast, and Cameroons. During this time, he also conducted research on the molluscan ecology of both coasts of Mexico and the Great Barrier Reef of Belize.

Continuing his education, Petuch studied marine biogeography and malacology under Gilbert Voss and Donald Moore at the Rosenstiel School of Marine and Atmospheric Sciences at the University of Miami, where he received a full scholarship. During this time, his doctoral dissertation research involved intensive field work in Costa Rica, Colombia, Venezuela, Barbados, the Grenadines, and Brazil, where he often went to sea with the local shrimpers for weeks at a time. After receiving his PhD in oceanography in 1980, Petuch was invited to conduct two years of postdoctoral research, funded by the National Science Foundation, with Geerat Vermeij at the University of Maryland. While there, he also held a research associateship with the Department of Paleobiology at the National Museum of Natural History, Smithsonian Institution, under the sponsorship of Thomas Waller, and conducted field work in the Plio-Pleistocene fossil beds of Florida and North Carolina and the Miocene fossil beds of Maryland and Virginia.

Petuch has also collected and studied living mollusks in Australia, Papua-New Guinea, Fijis, French Polynesia, Japan, the Bahamas, Nicaragua, and Uruguay. This research has led to the publication of over 350 scientific papers and the discovery and description of almost 2,000 new species of mollusks and almost 200 new genera. His previous 24 books are well-known reference texts in the malacological and paleontological communities, and some of the better known include: *Mollusks and Marine Environments of the Ten Thousand Islands* (2022), *Ancient Seas of Southern Florida: The Geology and Paleontology of the Everglades Region* (2021), *Tropical Marine Mollusks: An Illustrated Biogeographical Guide* (2020), *Jewels of the Everglades: The Fossil Cowries of Southern Florida* (2018), *The Living and Fossil Busycon Whelks: Iconic Mollusks of Eastern North America* (2015), *Cone Shells of the Okeechobean Sea* (2015), *Molluscan Communities of the Florida Keys and Adjacent Areas: Their Ecology and Biodiversity* (2014),

Biogeography and Biodiversity of Western Atlantic Mollusks (2013), *Molluscan Paleontology of the Chesapeake Miocene* (2010), *The Geology of the Everglades and Adjacent Areas* (2007), and *Cenozoic Seas: The View from Eastern North America* (2004). Currently, Petuch is a Professor Emeritus in the Department of Geosciences, Florida Atlantic University in Boca Raton, Florida where, for thirty years, he taught undergraduate classes in oceanography, paleontology, and physical geology, and graduate classes in paleoecology and paleoceanography. He currently resides in Jupiter, Florida, with his wife Linda, where they both enjoy visits from their three children and their families.

© Morgan Taylor Photography, with permission

David P. Berschauer was born in Rockville Center, New York, in 1964, and spent his youth collecting shells in such varied localities as New York, Florida, California, Washington, Mexico, and throughout the Caribbean. His early interests in natural history, malacology, and marine biology eventually led to a BS in biology at the University of California-Irvine, an advanced marine invertebrate zoology course at Washington State University's Friday Harbor Marine Lab, and studies towards the pursuit of a graduate degree in marine biology at Florida State University in Tallahassee, Florida. While still a college undergraduate, Berschauer performed field biology research, published a number of research papers and gave scientific presentations at national conferences. He subsequently switched career paths and attended Southwestern University School of Law in Los Angeles, California, earning his Juris Doctorate in 1991. Although having developed a legal career spanning over thirty years and being licensed in both California and Georgia, he has kept malacology as a lifetime avocation and has put together a sizeable research collection and personal museum of molluscan specimens.

Over his entire professional life, Berschauer continued to pursue his passion for marine biology, and collecting and studying marine organisms. In his spare time, he has developed and published a relational database software program to aid in the organization and maintenance of a systematic collection. Although originally designed for malacology, the program is applicable to entomology and other aspects of systematic zoology. Berschauer is an active member of the San Diego Shell Club, is a Museum Associate with the Natural History Museum of Los Angeles County ("LACM") Malacology Department, and is the Editor-in-Chief of the journal, *The Festivus*. He is also well known for his natural history and shell photography, with a multitude of high-quality examples seen throughout this book.

Besides being the author of many important scientific papers on molluscan systematics, Berschauer has described and named over 100 new species of gastropods and is also the co-author of several recent books on mollusks, including *The Living and Fossil Busycon Whelks: Iconic Mollusks of Eastern North America* (2015), *Jewels of the Everglades: The Fossil Cowries of Southern Florida* (2018), the well-received *Sea Shells of Southern California: Marine Shells of the Californian Province* (2018), *Tropical Marine Mollusks: An Illustrated Biogeographical Guide* (2020), *Ancient Seas of Southern Florida: The Geology and Paleontology of the Everglades Region* (2021), and *Mollusks and Marine Environments of the Ten Thousand Islands* (2022). He currently resides in Cairo, Georgia, with his wife Felicia, where they both enjoy visits with their children and their spouses, along with their young grandchildren.